Psychology of Problem Solving

BASIC TOPICS IN COGNITION SERIES
Sam Glucksberg, *Editor*

PSYCHOLOGY OF PROBLEM SOLVING

Theory and Practice

GARY A. DAVIS

BASIC BOOKS, INC., PUBLISHERS
NEW YORK

© 1973 by Basic Books, Inc.
Library of Congress Catalog Card Number 72-92795
SBN 465-06738-7
Manufactured in the United States of America

Based in part upon research and development conducted at the
Wisconsin Research and Development Center for Cognitive Learning

73 74 75 76   10 9 8 7 6 5 4 3 2 1

To Cathy

PREFACE

THE COMPELLING PROBLEM leading me to prepare this book was a very bothersome awareness of two immense bodies of information. Both highly developed sets of literature deal with the same psychological topic—human problem solving—yet in half a century the independent paths of these two ventures have seldom crossed.

On one hand is the chain of events appearing in psychology texts which link Wolfgang Köhler's (1925) classic and delightfully written studies of problem solving and intelligence in chimpanzees with modern bits of research and theory in experimental psychology. Some noteworthy scholars in this psychological tradition have been Thorndike (1911), Maier (1930, 1970), Katona (1940), Luchins and Luchins (1959), Wertheimer (1945); and more recently, Kendler and Kendler (1962), D. M. Johnson (1955, 1960), and in concept problem solving, Bourne (1966). Some exciting simulation work by computer scientists Newell, Shaw, and Simon (1958), Feigenbaum and Feldman (1963), and Hunt (1968, 1970), also must be included in our psychological sequence of problem-solving research and theory.

On the other hand is a problem-solving history evolving about the perpetual needs of industry for workable, innovative problem solutions. While the laboratory psychologist has been grappling with basic variables and processes of human problem solving, as is his bent, industry's venturous tack was to plunge in and teach some intuitively isolated attitudes and skills of creative problem solving. The earliest landmarks in this tradition are the creative problem-solving courses initiated by Crawford in 1931 (Crawford, 1937, 1954) and by General Electric in 1937. The person most directly responsible for the interest in training ingenuity is the late Alex Osborn (1963), developer in the 1930s of brainstorming and founder in 1954 of the Creative Education Foundation. Another leader is Gordon, whose *synectics* methods represent a problem-solving strategy approaching brainstorming in its impact (Gordon, 1961; Alexander, 1965). The concern with coaching skills of resourceful problem solving extends also into education, most notably and ably led by Parnes (1967), president of the Creative Education Foundation, and by Torrance (1962, 1965, 1970).

My design in writing this short book, then, was to summarize for the psychologist, the person in business or industry, and the educator, the sorts of research and speculation on problem solving and imagination that exist in his own area and in each of the other areas. While the book is in no way exhaustive, I have tried to describe some representative highlights from each perspective, showing commonalities where they exist, and, of more importance to everyone, exhibiting the theoretical and applied discoveries unique to each camp. In limiting the focus of the volume, some highly developed research topics necessarily have been ignored; for example, the topic of individual differences in problem solving and creativity (see, for example, Barron, 1969; Wallach and Kogan, 1965) and the very large literature on concept problem solving (Bourne, 1966).

Parts of this book, especially Chapter 10, are based upon research and developmental activities conducted during my tenure from 1965 to 1970 as Principal Investigator at the

Preface

Wisconsin Research and Development Center for Cognitive Learning, University of Wisconsin. I am grateful for the research support which I received during this period. I also wish to thank my delightfully bright crew of assistant-colleagues, Alice Train, Mary Manske, Sue Houtman, Bill Roweton, Tom Warren, Sami Mari, and Terry Belcher for adding their thinking to mine and for generally keeping me on my toes during these recent years. Finally, the excellent editorial suggestions by Sam Glucksberg are sincerely appreciated.

G.A.D.

CONTENTS

Part I	An Orientation	1
1	Introduction and Overview	3
2	Simplifying Assumptions, Definitions, and a Problem-Solving Taxonomy	12
Part II	Problem Solving in Experimental Psychology	27
3	Cognitive and Gestalt Views of Problem Solving: The Aware, Purposive Organism	29
4	Problem Solving to the *S-R* Psychologist: Transfer and Trial-and-Error	40
5	Computer Problem-Solving Models: A Tool to Match the Task	60
Part III	The Focus on Training: Industry and Education	75
6	Inquiry Activity: Study of the Compleat Problem-Solving Experience	79
7	Brainstorming	90

	8	Strategies for Stimulating Solutions: Attribute Listing, Morphological Synthesis, and Idea Checklists	103
	9	Metaphorical Thinking and Problem Solving: Synectics and Bionics	120
	10	Imaginative Problem Solving in the Classroom: The Wisconsin Project	133
	11	More Imagination and Problem Solving in the Classroom	150
Part IV		Summary	159
	12	Review and Comment	161

Appendix A	Tests and Measures of Creativity	165
Appendix B	Laboratory Problem-Solving Tasks	174
References		181
Index		199

◀ PART I ▶

An Orientation

IN OUR FIRST chapter we will sketch the two lines of thinking about problem solving and imagination: the *psychological* focus on understanding basic dynamics of problem solving and the *industrial/educational* emphasis on fostering skills of problem solving and innovation. Chapter 2 will define our most important concepts, make some assumptions and observations about problems and their solutions, and finally present a taxonomy of tasks used in problem-solving and creative thinking research.

◄ CHAPTER 1 ►

INTRODUCTION AND OVERVIEW

THE HUMAN ORGANISM is a continual decision-maker, problem-solver, and innovator. Many of life's little riddles fortunately require an inconsequential action, such as the daily decisions about what to wear and what to eat. On the other hand, many occasions scattered throughout our personal and professional lives demand more momentous decisions, where poorly considered solutions will be costly in dollars, happiness, or both. For example, we select a university and a professional career; we also reach decisions about marriage and family life. And in these educational, career, and family spheres, we regularly solve important problems and make major decisions affecting our lives and others' for years into the future.

On still a larger scale, we might observe that the history of civilization is itself a history of problem solving and innovation. Indeed, since the discovery of fire and the invention of the wheel, man's capacity for imagination and ingenuity has led to astronomical levels of convenience and scientific achievement. Imagine for a moment a heart-transplant patient watching color television as astronauts, transported by a giant Saturn rocket, step onto the moon. He listens as the space men describe their incredible voyage to the brownish-gray satellite, their voices and faces transmitted 239,000 miles from the moon to Houston, Texas. Of course, not only medicine and rocketry demonstrate imaginative problem solving. One has only to thumb through any issue of *Time* or *Newsweek* to witness up-to-the-minute innovations in the literary, performing, and graphic arts, in education, always in consumer products and their promotions, sometimes in the ministry, and all too often in military gadgetry.

In view of the conspicuous importance of human problem solving and creativity, then, it is not surprising that psychologists and many others have devoted large portions of their professional lives to unraveling these fascinating behaviors. While the mysteries of creative problem solving—like human consciousness itself—probably will not be totally understood in this century, many inroads have been made into the psychological and environmental conditions that promote problem solving and creative behavior.

Overview of This Book

The strategy of this book is to examine two sets of literature dealing with human creative problem solving. From the field of psychology, we find a body of knowledge seeking to describe and understand human problem solving in the laboratory and in the real world. From education and industry, taken together, we find a legitimate interest in the systematic training of attitudes, abilities, and skills of creative thinking and problem solving. Somehow, for decades, people from each of these two main approaches have made great strides toward understanding human problem solving; yet strangely enough, each group has rarely consulted or acknowledged the literature of the other field.

Following the early leads of Wolfgang Köhler and E. L. Thorndike, the modern experimental psychologist continues to identify and describe (1) task variables influencing problem difficulty, (2) thinking styles and strategies, and (3) personality traits that help or hinder effective problem solving. His laboratory tasks are necessarily contrived and artificial in the attempt to study human problem solving in its uncontaminated, purest form. For example, the most often used laboratory problem is *anagrams,* a task requiring the human subject (*S*) to rearrange scrambled letters into a meaningful word. Even anagrams, a conceptually simple task in the sense that blind trial-and-error can eventually produce the solution, has helped to clarify such general factors as the effects of language habits upon problem solving.[1] The anagram task also has been used as a vehicle to study such problem-solving phenomena as *fixation* (the tendency to continue responding as trained despite the fact that a shortcut is available) and *learning set* (the improvement in problem solving and learning which occurs over a series of similar tasks).

There are, of course, a great many other laboratory problems.[2] Always, however, the experimental psychologist seeks to identify and evaluate basic or "pure" problem-solving variables, to examine repetitive problem-solving phenomena and behaviors, and occasionally to develop a theory of human problem solving. Some of these theories are general speculations on the nature of human mental life. For example, Köhler used observations from his chimp studies to help formulate his *perceptual reorganization* (Gestalt) viewpoints on thinking. Also, Hull (1934) developed the *habit-hierarchy* concept of thinking and problem solving: the notion—based partly on rat research—that a stimulus situation will elicit first the strongest habits and later the weaker responses, strength being determined by the organism's history of reinforcement. Other problem-solving theories are task-specific, as when Ronning (1965) formulated a *rule-out* explanation of anagram solving, based on the notion that the solver rules out unlikely letter combinations (such as QL).

Part II of this volume will examine three branches of psychological research and theory in problem solving. In Chapter 3 we will take a closer look at classic and recent cognitive-Gestalt theories: for example, Köhler's famous chimpanzee studies and Bruner's concept of higher-order codes. Emphasizing principles of learning and conditioning instead of the Gestalt insight, Chapter 4 will examine Thorndike's classic trial-and-error studies of problem solving, followed by theorists who emphasize other experience-based mechanisms, such as habit-hierarchies, transfer, operant conditioning, and response chaining. The newest psychological branch, one originating in the field of computer sciences, is the computer-simulation of problem solving described in Chapter 5. Neither the cognitive-Gestalt theorist nor the learning psychologist has ever specified in such detail the steps and information processes that the computer expert must clarify in order for his machine to duplicate human problem-solving behavior.

No doubt, interests and values will flourish where they are highly rewarded. It is in industry, then, whose profit and virtual survival depend upon ingenuity and resourcefulness, where creative problem solving has been scrutinized with a view toward teaching some useful principles and strategies for creative behavior. The first professional course for training creative problem solving was initiated in 1931 by the late Robert Crawford (1954), Professor of Journalism at the University of Nebraska. While Crawford was formulating his principles of *attribute listing* (Chapter 8), General Electric launched its problem-solving course in 1937. However, the lion's share of the credit for early interest in training creative problem solving belongs to Alex Osborn, cofounder of the New York advertising agency Batten, Barton, Durstine & Osborn, founder in 1954 of the Creative Education Foundation, and creator of the brainstorming procedure (Chapter

7). Osborn's early books, *How to "Think Up"* (1942), *Your Creative Power* (1948), and *Wake Up Your Mind* (1952), and the early editions of *Applied Imagination* (1963), influenced the development of a great many creative problem-solving programs in industry. Since these early years, virtually every large corporation in America conducts creative problem-solving courses for executives, lower management, and especially their valued engineers. A partial list includes A. C. Spark Plug, American Telephone and Telegraph, Dow Chemical, Eastman Kodak, B. F. Goodrich, IBM, Sylvania Electric, United Airlines, RCA-Whirlpool, Xerox, U. S. Steel, and the U. S. Army, Air Force, Navy, and Marine Corps. Many university business and engineering departments conduct similar creative problem-solving courses for potential members of the same commercial audience.[3]

Several corporations deal with nothing but creative problem-solving, its analysis and training. Synectics, Inc., whose colorful background and activities are described in Chapter 9, may claim at least three major functions: it trains creative thinking groups primarily for large corporations; it solves tricky engineering problems for clients (nicely practicing its own preachings); and, recently, Synectics founder Gordon has been writing and testing educational materials for inner-city schools. Harbridge House, Inc., also thrives on commercial creative problem solving, largely through its *value engineering* services. Value engineering, a form of creative cost cutting, is taught and practiced by most automobile manufacturers (Ford, for example) and aircraft companies (Lockheed, for one). Most individual value engineers are members of the Society of American Value Engineers—SAVE—a society committed to the importance of applying systematic problem solving to all phases of product development and marketing.[4]

While industry and education seem to differ in most of their major goals, they are rather united in their common interest in tutoring skills of effective and imaginative problem solving. However, major developments in education have been noticeably slower than in the business world. In fact, even though many flexible teachers improvise their own strategies for stimulating thinking, problem solving, or creative experiences, only a handful of theory-based and field-tested educational materials exist. Of these, a pitiful number are commercially available for wide distribution.

Part III of this book, then, will summarize a number of professional methods for finding creative new ideas and problem solutions—brainstorming, attribute listing, using idea checklists, the metaphorical synectics methods, plus the biology-based engineering strategy of bionics. From education we will look at the inquiry-training approach to skill-development, along with other educational strategies and materials for stretching young imaginations.

Creative Problem-Solving Model: Attitudes, Abilities, and Techniques

To simplify the content of our various professional and educational creative problem-solving courses and materials, I have found it helpful to keep in mind three main components of this training: *attitudes* conducive to creative behavior, basic *abilities* underlying creative potential, and deliberate *techniques* for the production of new ideas and idea-combinations. This three-part model helps unscramble the nature of "creativity" and creative problem solving, and serves to clarify what may be taught when we "teach creative problem solving."

While a more complete discussion of creative attitudes, abilities, and techniques appears in Chapters 10 and 11, we may now elaborate just a little. Creative *attitudes* include a conscious intent, sometimes even eagerness, to search for imaginative ideas and viewpoints, to escape from habit, tradition, and conformity pressure, and to seek novel products and problem solutions. Such attitudes are essential for flexible thinking and thus lie at the core of every creative problem-solving course or program, from Torrance's elementary school exercises (Torrance, 1970) to Lockheed-Georgia's (1971) courses for engineering design. We will argue in Chapter 10 that a major portion of the apparently mysterious creative potential is quite voluntary, determined directly by the individual's attitudes toward new ideas, change, and innovation.

Creative *abilities* refers very broadly to those mental capabilities necessary for problem solving and creativity: for example, abstracting, combining, perceiving novel relationships, associating, imagining, filling in missing information, transforming, or taking an "intuitive leap." While it is difficult to speak of training such basic mental abilities, nonetheless several of Torrance's educational programs described in Chapter 11 seek to strengthen such abilities through exercise and reinforcement.

Techniques for creative thinking are procedures for producing new ideas consciously and deliberately—without waiting for an unpredictable inspiration. Actually, most of our creative thinking techniques, such as brainstorming, attribute listing, checklisting, and the synectics methods, were initially derived from the introspective reports of creative individuals. The procedures are, in a sense, "unconscious" methods which have been specified and made operational. We normally would use such "forced" techniques to supplement our intuitive supply of problem solutions.

Scope

It is my intent in this short book to be as comprehensive as possible. Whether solving anagrams in the psychological laboratory, brainstorming in industry or the classroom, or solving everyday problems in the real world, there is sufficient commonality that some clarifying principles—such as the interfering role of habit—apply easily to all problem situations. Certainly, noting the continuity between seemingly unrelated forms of problem solving or creativity should help clear away some conceptual and definitional cobwebs. At the same time, it also is true that many problem-solving tasks and creative thinking situations are distinctly different, not only in content area but in the strategies and solution processes involved. To best understand human creative problem solving, then, we also must try to identify and acknowledge these important differences.

Summary

The critically important yet elusive topics of human problem solving and creativity have earned the serious attention of a great many psychologists, educators, and individuals in business and industry. This book will describe and draw concepts from major developments within two independent traditions, the psychological and the industrial/educational. The goal is to summarize major efforts in these areas, to clarify the nature of human problem solving and creativity as it occurs in the laboratory or real world, and to identify and describe reasonable principles for improving problem solving and creative skills. Our three-part model, which conceptualizes an effective, creative problem-solver as possessing certain attitudes, abilities, and techniques, may aid in achieving some of these goals.

NOTES

1. For example, the "meaningfulness" of the solution word, the frequency of occurrence of two-letter and three-letter configurations in the English language, the pronunciability of the anagram, and the similarity of the anagram to the solution word.
2. A list of laboratory problem-solving tasks, each with a brief description, appears in Appendix B.
3. See Edwards (1968) for a more complete listing of creative problem-solving courses in various industries and colleges.
4. For an overview of the Lockheed-Georgia Value Engineering Department, see their chapter in Davis and Scott (1971).

◄ CHAPTER 2 ►

SIMPLIFYING ASSUMPTIONS, DEFINITIONS, AND A PROBLEM-SOLVING TAXONOMY

HUMAN PROBLEM SOLVING and innovative thinking are among the highest and most complex forms of human mental life. Such thinking takes innumerable shapes and forms in a wide variety of problem-solving and decision-making situations. Given this broad subject matter, it would be useful to define some pivotal terms, make a few assumptions and observations, and finally try to classify various types of problem situations. The goal throughout is to specify both the commonalities and distinctions among, for example, Köhler's chimpanzee Sultan raking in a banana; a college student solving an anagram, chess, bridge, or crossword problem; or a mechanical engineer creating a low-pollutant steam engine. Some definitions and assumptions will be "safe": that is, obvious or circular. Others may not be and perhaps will not stimulate instant agreement by the thoughtful reader.

A Problem

A simple yet meaningful definition of a *problem* is as follows: *A problem is a stimulus situation for which an organism does not have a ready response.* Skinner (1966) pro-

Simplifying Assumptions and Definitions 13

posed a similar definition: "A question for which there is at the moment no answer is a problem." Since "questions" and "answers" imply verbal behavior, and since much problem solving is nonverbal (consider problem solving in music, in mathematics, in art, and by animals), the former definition is the more general and preferred one.

As with other seemingly simple and unambiguous definitions, however, closer analysis easily muddies the waters. We must further assume, as did Hoffman (1961), that the organism is motivated or constrained to find a solution. Shulman (1967), following Dewey (1938), defined a problem as a "psychological state of discomfort or disequilibrium as sensed by an individual" (see Chapter 6). Both Hoffman and Shulman, then, would agree that our definition of a "problem" as a stimulus situation for which an organism does not have a ready response, cannot leave the problem-solver the option of avoiding or ignoring the "problem."

It follows from our definition that a "problem" for a naïve organism may not be a "problem" at all for a more sophisticated organism. The latter *does* have a ready response for the same stimulus situation which confuses the former. Strictly speaking, our definition further implies that once we solve a given problem, we then should call it something else. Unfortunately, our language does not provide an alternate label for a problem whose solution is well known, such as the arithmetic "problem" of $2 + 2$.

These considerations aside, our definition of a problem— a stimulus situation for which the organism does not have a ready response—is sufficiently general to apply to all of the laboratory and real-world creative problem solving that we will describe.

The Solution

Our knowledge of problem solving and creativity will not be enhanced if we define a *problem solution,* as do some scholars, as "that which solves the problem." Equally unenlightening is, "that which meets the requirements of the problem." There are also more poetic, yet equally meaningless definitions which provide a misleading impression of authoritativeness. Consider: ". . . [creative problem solving] is the encounter of the intensively conscious human being with his world" or ". . . [creative problem solving] is the ability to *see* and to *respond.*" And how about: "It [creative problem solving] involves acting as a total person, with subconscious and unconscious levels acting in some form of unity with conscious levels."

We will define a *problem solution,* or a *creative idea,* as a new combination of existing ideas. Such a definition is not original; it is, in fact, the most common among hundreds of definitions.[1] No doubt the reason for the popularity of this simplifying definition is that it has several helpful implications. First of all, considering a "new" solution to be a novel combination or rearrangement of existing ideas is far more meaningful and comprehensible than attributing the solution to insight or to the unconscious. Note also that the present definition points to the critical role of a broad knowledge-base in problem solving, for without the requisite component ideas, the appropriate solution, or idea combination, is rather unlikely to appear.

The definition of a new idea or problem solution as a unique combination of ideas serves also to translate theory into the training of creative problem solving. We readily understand combining and rearranging. And as we will see in Part II, we can teach strategies for producing new idea com-

Simplifying Assumptions and Definitions

binations, thereby increasing one's capacity to produce creative problem solutions.

Finally, our definition of a *solution,* as with our definition of a problem, has generality across many qualitatively different types of problem situations. For example, anagram solving clearly requires a letter recombination; engineers, artists, and clothes designers combine colors, materials, and forms to produce new products; and a mathematician combines existing ideas into creative mathematical solutions.[2]

Steps and Stages in Problem Solving

Often in psychology, when we do not understand a mental (hence unseen) activity, we call it a process. Thus we see the phrases "learning process," "memory processes," and even "information-processing processes." Over the years, several scholars have sought to shed light upon the problem-solving process by specifying in detail the sequence of events—the steps and stages—through which one might proceed in solving a problem.

It is self-evident that problem solving minimally requires that we first become aware of the problem, then proceed to solve it. All of the following stage-analyses are elaborations of this truism.

Best known are the stages listed by Wallas (1926): (1) *preparation,* composed mainly of clarifying and defining the problem, along with gathering pertinent information; (2) *incubation,* a period of unconscious mental activity assumed to take place while the individual is (perhaps deliberately) doing something else; (3) *inspiration,* the "Aha!" or "Eureka!" experience, which occurs suddenly; and (4) *verification,* the checking of the solution. It should be noted that these steps correspond closely to steps in the scientific method: state the problem, propose hypotheses, plan and conduct research, and finally, evaluate the results.

The stages of incubation and inspiration, by definition, are unobservable mental events. However, both concepts are real, albeit elusive elements in human problem solving and thinking familiar to anyone who has unexpectedly had a "good idea" in response to yesterday's problem. The same psychologists who doubt the scientific status of incubation or illumination also keep a pad and pencil on the bedstand for just such happenings. Well-documented introspective reports of great discovery by "incubation" followed by "inspiration" are too common to ignore. For example, two of the more illustrious writers on creative mathematical problem solving, Polya (1957) and Hadamard (1945), along with other nonpsychologist experts on creative problem solving (Gordon, 1961; Mason, 1960; Osborn, 1963), readily acknowledge the unconscious, preconscious, or sometimes "fringe-conscious" activity leading to the solution of a difficult problem. Unfortunately, the precise mechanics of incubation and inspiration are rather difficult to isolate, since we would be faced with studying phenomena of which, by definition, we cannot even be aware!

Perhaps to the relief of nonbelievers, some stage-analyses of problem solving skip the incubation step entirely. The steps of Kingsley and Garry (1957) include: (1) a difficulty is felt; (2) the problem is clarified and defined; (3) a search for clues is made; (4) various suggestions appear and are tried out; (5) a suggested solution is accepted (or the thinker gives up); and (6) the solution is tested. Dewey (1933), writing on the nature of thinking, seemed to summarize the Kingsley-Garry six steps in just two, with no loss of information: first, there appears a state of doubt, perplexity, or mental difficulty in which the thinking originates, followed by an act of searching, hunting, or inquiring to find material that will resolve the doubt, and settle and dispose of the perplexity.

Simplifying Assumptions and Definitions 17

A final stage-formulation we should acknowledge is by the late Alex F. Osborn of brainstorming fame. In Chapter 7 we will see that Osborn's single most important principle of creative problem solving is the distinction between idea-*creation* and idea-*evaluation*. The following ten steps show Osborn's recommended alternations between creative (steps 1, 3, 5, 7, and 9) and evaluative thinking (steps 2, 4, 6, 8, and 10): (1) think up all phases of the problem; (2) select the subproblems to be attacked; (3) think up what data might help; (4) select the most likely sources of data; (5) dream up all possible ideas as keys to the problem; (6) select the ideas most likely to lead to solution; (7) think up all possible ways to test; (8) select the soundest ways to test; (9) imagine all possible contingencies; and (10) decide on the final answer (Osborn, 1963, pp. 207–208).

In sum, it is the ostensible purpose of these analyses of steps and stages to simplify, explain, and—in Osborn's case—recommend the sequence of events in successful problem solving. These simplifications are helpful if we keep some qualifications in mind. Whether solving an anagram or designing a generator, a person solving a problem will *overlap his steps,* as when he begins to think of solutions while defining the problem; he will *backtrack to earlier stages,* as when he finds it necessary to redefine or further clarify the "givens"; and he may *skip some steps* entirely, especially the *incubation* one.

Creative Problem-Solving Skill: Intuitive or Learned?

As with any other human characteristic, mental or physical, creative and problem-solving abilities—whatever they are—are distributed unevenly. We have Werner von Brauns, Pearl Bucks, Duke Ellingtons, and Pablo Picassos, who possess obvious creative genius, and we also have less resourceful professional and amateur scientists, writers, musicians, and artists. The profound discovery of every psychology graduate student is the incredible variation among his human experimental subjects in any trait or ability measured. Even Wolfgang Köhler's chimpanzee Sultan was credited with being the brightest of Köhler's primate problem-solvers.

It is impossible to determine whether these differences are due entirely to heredity, entirely to learning, or—most likely—to both. In any event, the skills involved can be increased through the deliberate teaching of appropriate attitudes related to creative problem solving and techniques for producing ideas, and by exercising various abilities. Hopefully, the skeptical reader will withhold final judgment until reading later chapters, which describe in detail the "what" and "how" of training imagination in industry and education.

Naturally, heredity, like experience, will place limits on creative achievement, just as these two factors limit all other facets of intellectual and physical development. However, research and professional experience have shown that gains in creative problem-solving attitudes and abilities are easily substantial enough to justify the increasing industrial and educational interest in their training.

Barriers to Effective Problem Solving: Habit and Conformity

Aside from a lack of prerequisite information or deficient mental gymnastic ability, there are two main interrelated barriers to effective problem solving: *habit* and *conformity* pressure. Habit and conformity are implicit in such key problem-solving and personality concepts as rigidity, fixation, mental set, perceptual set, predisposition, resistance to change, tradition-orientation, fear of the unknown, and on occasion, pigheadedness. We will see in Chapter 3 that the core of Gestalt psychological problem-solving theory lies in

Simplifying Assumptions and Definitions 19

overcoming *fixation,* which means resisting the incorrect but habitual (usual, normal, shortest) route to the goal.

All psychological experiments with water-jar problems try to show how easily normal humans can become fixated on a wrong solution: after carefully learning a formula for producing the required volume of water (Jar $B-$ Jar $A-2$ Jar Cs) and applying it successfully to five problems in a row, the average person will not see a more direct, simpler solution (Jar $A-$ Jar C) to the next problem. Such fixation is nothing more than a quickly conditioned habit, reinforced by a series of successes.

In industrial creative problem-solving programs, the first order of business is to change attitudes from resistance to receptiveness to innovation and change. In his fine book on brainstorming, Clark (1958) listed some all-too-familiar "idea-squelchers," most of which directly reveal habit and conformity: "It has been the same for twenty years so it must be good." "We can't do it under the regulations." "We've never used that approach before." "It's not in the manual." "Our people won't accept it." "We're not ready for it yet." "What will the customers think?" "Somebody would have suggested it before if it were any good." "Too modern." "We have too many projects now." "I just know it won't work." "Production won't accept it." "They'll think we're longhaired." "You'll never sell that to management." "Don't move too fast." "Why something new now?" "Our sales are still going up." "The union will scream." "Here we go again." "I don't see the connection." "Won't work in our industry." "Political dynamite." "It's not in the plan." "It's too early." "You don't understand the problem." "No adolescent is going to tell me how to run my business."

Types of Problems and Problem Solving: A Taxonomy

We have already noted that human problem solving and thinking are highly complex and may assume innumerable specific forms. A survey of the laboratory tasks and creativity tests described in Appendixes A and B would show that, just considering artificially devised problems, the diversity of problems and their elicited activities is tremendous. Some tasks are highly verbal: for example, anagrams, tasks requiring someone to produce words with specified first and last letters (first and last letters test), or tasks requiring the production of a response word given the stimulus words (associations test). Some problems are mathematical, such as water-jar problems or asking S to predict the next number in a series (number series problems). Some tasks are mechanical, as when S is required to construct a candle-holder (candle problem) or a pendulum (pendulum problem). Many problems demand a trial-and-error approach, as in learning to classify figural patterns (concept problems) or trying to achieve a particular light pattern by manipulating switches (switch-light problems); others require implicit problem solving and thinking: for example, thinking of ways to destroy a tumor with X-rays without damaging surrounding tissue (X-ray problem). Many laboratory tasks are highly structured, providing all that is needed to reach a solution, while others are deliberately open-ended.

Obviously, it would be a serious mistake to casually conclude that all problem solving is basically the same. One may reach this conclusion, but it should not be reached casually. For example, Jacques Hadamard (1945) noted, "It is obvious that invention or discovery, be it in mathematics or anywhere else, takes place by combining ideas." Gerard (1952) also observed, "Many have insisted that the imagina-

Simplifying Assumptions and Definitions   21

tion process is different in art and science. . . . On the contrary, the creative act of the mind is alike in both cases." At the other extreme is the position of Fox (cited by Fabun, 1968), who claimed, "It would seem quite apparent that there is no *one* creative process, and there may well be as many creative processes as there are creative people."

In any event, there are several important and identifiable dimensions of problems and problem solving that allow us to recognize points of commonality among a great many forms of laboratory and real-world problem solving. Furthermore, these dimensions, which I will cast in the form of four questions, allow us to distinguish between tasks which, on the surface, seem identical. The questions:

1. Is the problem really a problem?
2. Does the task elicit observable trial-and-error behavior or implicit problem solving and thinking?
3. Does the task require one "correct" solution or many original ones?
4. Is the problem (and its solution) a fairly well-defined, one-shot affair, or is it a creative contribution of substantial magnitude, requiring the creative solving of multiple subproblems?

The first question—"Is the problem really a problem?"—allows us to identify situations that simply do not meet our definition of a problem as a *stimulus situation for which the organism does not have a ready response*. Simple arithmetic "problems" (addition, making change) and simple questions ("Who invented the Ford?") are not "problems" at all to average adults. Unnecessary definitional confusion results from trying to include these nonproblems when discussing tasks that do meet our definition of a problem.

The second question—"Does the task elicit observable trial-and-error behavior or implicit problem solving and thinking?"—provided a useful basis for an earlier problem-solving taxonomy (Davis, 1966). A large number of seemingly unrelated problems stimulate good old-fashioned and highly effective trial-and-error searching. Switch-light tasks, concept problems, and other laboratory tasks, along with such real-world difficulties as hunting for a lost checkbook, troubleshooting a mechanical bugaboo, or inventing a new recipe, require substantial trial-and-error searching for a good solution. On the other hand, a perhaps larger variety of situations stimulate implicit problem solving. For example, we mentally unscramble an anagram problem, "visualize" alternatives to the X-ray problem, or thoughtfully consider different ways to organize a term paper.

The third question—"Does the task require one correct solution or many original ones?"—calls attention to a glaring distinction between traditional laboratory tasks—anagrams and the pendulum, X-ray, switch-light, water-jar, and concept problems—which always require one correct answer, and tasks that require a large number of original ideas or solutions. With the latter, for example, one may be asked to list uses for an automobile tire (unusual uses test) or to think of improvements for a stuffed toy animal (product improvement test).

We must caution, however, that this distinction, which has been referred to as "divergent thinking" (producing many solutions) versus "convergent thinking" (finding one solution), may be misleading in the context of creative problem solving in the real world. For example, the engineer groping for a design for a better electric generator will first "divergently" think of a number of possibilities; but he likely will "convergently" select and implement only the solution that best meets his perception of problem requirements. Both types of thinking are likely to appear in the same problem-solving episode.

Finally, the fourth question—"Is the problem (and its solution) a fairly well-defined, one-shot affair, or is it a creative contribution of substantial magnitude, requiring the creative solving of multiple subproblems?"—simply acknowledges an obvious distinction. Such "single" problems as landing a man on the moon (and getting him back) or composing a symphony, would fall at one extreme of this complexity dimension, while any laboratory or other short-term problem would lie at the other extreme.

By keeping these fairly straightforward questions in mind as we study problem solving, we will find that regardless of subject area—science, music, mathematics, chess-playing, laboratory problems, or daily personal problems—commonalities exist. For example, many simple responses to stimuli need not be considered "problems" at all, according to Question 1. Question 2 warns us to look for familiar trial-and-error problem solving—or learning—the most basic and pervasive solution-activity of all.

The questions also help identify subtle, but important, distinctions between seemingly identical forms of problem solving. Anagram problems actually appear in three different forms, eliciting three fairly distinct types of solution-activities. In the typical case, the person must mentally rearrange the problem letters to find the solution word. But if a pencil and paper or "alphabet blocks" are provided, much of his anagram-solving activity will become overt trial-and-error. Also, there exists a multiple-solution anagram task, requiring the individual to create as many words as he can from the letters in a given stimulus word (say, *factory*). Thus the multiple-solution problem stimulates a third variety of anagram-solving as described by the one-versus-many distinction of Question 3.

While the implicitly solved anagram problem may become an overt trial-and-error task if S has the equipment, a trial-and-error switch-light problem will become an implicit task if S is properly pretrained. In an earlier experiment, the author (Davis, 1965, 1967) discovered that college students who first memorized all switch-light connections could produce the required light pattern by implicitly deducing the solution. Untrained subjects, who could not predict the outcomes of the response alternatives (switches), necessarily used overt trial-and-error. Other problem situations may also elicit overt trial-and-error at one point and implicit activity at another, depending on whether S can predict or "foresee" the success-value of his available solution-alternatives.

Summary

I have presented some simplifying assumptions, definitions, and observations regarding the nature of problems and their solutions, along with a taxonomy of problem-solving tasks. (1) A *problem* was defined broadly as a stimulus situation for which an organism does not have a ready response. Given that the organism is motivated or constrained to remove the disequilibrium aroused by the problem, such a broad definition encompasses virtually any form of problem solving from matchstick puzzles to a student faced with scraping up tuition fees. (2) I defined a *problem solution,* or a *creative idea,* as some new combination of existing ideas. This definition simplifies the mysterious nature of a "new idea" and—like my first definition—points to an important similarity between various thinking and problem-solving situations. (3) More as an observation than a definition or assumption, I noted that problem-solving behavior seems to follow the minimal two steps of detecting, then solving the difficulty. Others have sought to describe the problem-solving process by reference to such steps and stages as *preparation, incubation, illumination,* and *verification.* (4) I proposed that, within limits, creative problem-solving skill can be improved by teaching attitudes and idea-producing techniques, and by exercising basic abilities. (5) Habit and conformity, in the form of fixation, set, rigidity, and tradition-orientation, were described as the critical barriers to problem solving and innovation. (6) Finally, I described a taxonomy of problems primarily based on the type of elicited activities. The goal was to reduce the incredible variety of problem situations across diverse subject-areas by calling attention to four dimensions: Is it really a problem? Does the task elicit overt trial-and-error or implicit behavior? Does the solver provide one or many solutions? Is it a brief problem-solving episode, or a large-scale project containing many subproblems?

We also noted that the distinction between "convergent" and "divergent" thinking—which comprises the basis of the third question above—may be misleading. The convergent problem-solver may divergently produce many alternatives before finding the "right" one; and the divergent thinker can, if requested, convergently evaluate and present his single best idea.

NOTES

1. See, for example, the numerous definitions appearing in Fabun (1968) and throughout Taylor and Barron (1963) and Parnes and Harding (1962).
2. My personal contacts with art, engineering, and mathematics professors have indicated that such a definition is meaningful, acceptable, and already quite familiar to each of them.

◀ PART II ▶

Problem Solving in Experimental Psychology

TRADITIONALLY, the major theoretical schism in psychology has been the near century-old dispute between S-R behaviorists and Gestalt or other "cognitive" theorists. Many of the basic arguments between the two schools lie in their "mechanistic" versus "mentalistic" views of human problem solving.

To the cognitive-Gestalt psychologist, a problem by its nature disturbs the equilibrium of the organism. Solving the problem, which restores psychical balance, is a matter of achieving "insight," a construct also known as cognitive restructuring, perceptual reorganization, illumination, or recentering. In Chapter 3 we will scan the highlights of the classic primate work of Wolfgang Köhler, early German Gestaltist. Brief reference also will be made to cognitive-Gestalters Luchins, Maier, and Sheerer. A recent and different form of cognitive psychology will be represented by Bruner, who uses the word *coding* to describe the categorizing, principle-learning, and theory-construction that enables the conscious human to solve many kinds of problems.

To the archetypical behaviorist, the only evidence of *insight,* the pivotal Gestalt concept, is that the thinker visibly solves the problem. Hence, by substitution, "solving the problem by achieving insight" logically reduces to "solving the problem by solving the problem," and little is gained through the use of such terms as *insight, recentering, mind,* and the like. The learning psychologist, past and present, seems to have a choice of basically two main problem-solving explanations: (1) the person may transfer prelearned experiences to the new problem situation, or (2) he may proceed by trial-and-error, as in any other instrumental learning situation. Either way, the problem-solving response is learned, and as such, obeys the same laws of conditioning as any other learned behavior. Chapter 4, again beginning fifty years ago, describes the classic trial-and-error view of Thorndike, followed by more contemporary learning-based interpretations of problem solving offered by Staats, Skinner, and others.

Computer-simulation of problem solving (Chapter 5), pioneered by MIT computer scientists Newell, Shaw, Feigenbaum, and Simon, mostly ignores the longstanding learning versus cognitive doctrines. The strategy is to use computer programming to determine when the sequence of problem-solving activities is described in sufficient detail to allow a machine (and by inference, a man) actually to solve a given problem. At times, the computer will test and reject solution-alternatives in a simple trial-and-error fashion; at other times, it will appear to make inferences and logical deductions or follow complex "cognitive" rules.

◄ CHAPTER 3 ►

COGNITIVE AND GESTALT VIEWS OF PROBLEM SOLVING: THE AWARE, PURPOSIVE ORGANISM

THE DISTINGUISHING CHARACTERISTIC of the Gestalt and the more contemporary cognitive psychologist is his commitment to explain human behavior in its everyday, conscious, and strategic purposiveness. The theoretical language of the cognitive-Gestalt theorist reflects his phenomenological level of study: the organism perceives, thinks about, and analyzes his environment; he forms tenable hypotheses, tries plausible leads, follows rules, reasons, encodes, deduces, and makes predictions and calculated guesses. Such "mentalistic" concepts contrast sharply with the stimulus-response-reinforcement language of the learning psychologist.

In this chapter, we intend to present the flavor of traditional Gestalt theory as it applies to problem solving by describing briefly the classic research and observations of Wolfgang Köhler. Then, following a few comments on the role of fixation, set, and direction by cognitive-Gestaltists Luchins, Sheerer, and Maier, some of Jerome Bruner's ideas will represent a noticeably different, contemporary flavor of cognitive problem-solving theory.

Wolfgang Köhler

Köhler's monumental work *Mentality of Apes* (1925) summarizes his four active years studying primate problem solving at the Anthropoid Station on Tenerife, largest of the British-held Canary Islands where, as a German, Köhler was stranded for the duration of World War I. To Köhler a problem is said to exist when the direct route to a goal is blocked, resulting in psychic tension. Not accidentally, every one of the problems Köhler devised were called *roundabout way* tests. For example, in his detour problem (solved by chimpanzees, one female dog, a three-and-a-half-year-old girl, and not too successfully by several hens), Köhler placed the goal object just on the other side of a wire fence, in easy view of the subject. Reaching the goal required a 180° turn and movement away from the goal, around the U-shaped barricade to the treat.

To Köhler, the process involved in a true problem solution, not to be confused with an accidental success, must include the key Gestalt concept of insight: a cognitive reorganization or restructuring in which a nondirect path is perceived as leading to the tension-reducing goal. This "genuine achievement" occurs suddenly, marked in fact by a quick change in behavior: the dog stops and quickly spins around, or the child's face suddenly lights up. Following the insight, a smooth continuous action achieves the goal object.

Using bananas for rewards, Köhler tested the chimps' wits and observed quick perceptual reorganizations in a number of often-frustrating problems. In one group of tasks, all of which involved a banana or basket of bananas suspended out of reach from the ceiling, the apes learned to (1) release the free end of the string from a tree, thus dropping the basket; (2) place a bamboo pole vertically under the banana, and instantly climb up to twelve feet before the pole fell; and (3) push a box (or stack several boxes) under the bananas to use as a footstool. In one creative version of the latter solution, chimp Sultan pulled Köhler under the banana, climbing the famed psychologist as he would a ladder. In still another variation, the chimps learned to use each other as footstools. Unfortunately, with increasing sophistication, all wanted to be climbers and none stools, resulting in noisy battles among the chimpanzees.

The task for which Köhler and his chimps are best known is the stick problem. With a banana carefully placed beyond the chimp's reach, the subject must use a stick to rake in the prize. The direct goal path was simply stretching for the banana with those long arms, or when chained by the neck, with a farther-reaching foot. After varying amounts of energetic arm, shoulder, and foot stretching, the chimp would suddenly (read: "insightfully") seize the stick and adroitly drag in the banana. Once in a while, however, he would push the prize on the wrong side, putting it out of distance for good.

Köhler devised several challenging variations of the stick problem. One version involved a detour-type arrangement. The banana was placed in a three-sided drawerlike box, with the open side away from the cage. The ape first had to push the banana away from himself, out of the drawer, angle it past the side of the box, and then shuffle it in. Another version required the ape to push a large box out of the way in order to reach the banana with a short stick.

The most difficult task of all was the two-stick problem, to which Sultan, the most famous of the chimps, lent his talents. To reach the banana, the smaller bamboo stick had to be fitted into the end of a slightly larger one, creating a single stick long enough to reach the distant goal object. For the first fruitless hour, Sultan committed "bad" errors, such as dragging a box from the rear of his cage and placing it directly in his own way, and "good" errors, pushing one stick with the other toward the banana. Finally, ignoring the perturbing banana, Sultan played carelessly with the two sticks; and when he accidentally slipped the little one neatly into the bigger one, the light blinked on and he instantly dashed to rake in his dinner.

In addition to the many instances of insight, Köhler further recorded complex forms of primate reasoning, including implicit trial-and-error behavior. For example, Sultan would not mount a box too small nor thrust a stick too short to reach the banana, implicitly predicting the futile outcomes.

Another problem-solving facet of ape mentality was that of their tool-making and tool-using capabilities. For example, one of the chimps' fondest amusements was thrusting a pointed stick (which they had sharpened) at spectators, dogs, and especially hens. Fowl-stabbing developed to the point where one chimp baited the hens by dangling a piece of bread through the bars so that his scurrilous partner could jab the feathered body of the ever-unsuspecting chicken.

Regarding favorable conditions for problem solving—that is, for achieving insight—Köhler's single emphasis throughout his volume was upon the perceptual configuration: the Gestalt. In both the detour and stick problems, the insight always occurred under favorable "optical" or "geometric" circumstances which, to Köhler, were absolutely indispensable. The chimps solved the detour problem only when they achieved an adequate view of the possible indirect routes. With the stick problem, the critical spatial configuration determined the success of even experienced chimps: if the stick were not visible when looking at the banana, and vice versa, if a look at the stick removed the banana from view, the ape almost always was unable to use the stick effectively. He had to see the objects "in connection."

Köhler's very strong emphasis on the perceptual or geometrical configuration is consistent with our own description of original problem solving. In Chapter 2, we defined a

Cognitive and Gestalt Views 33

problem solution as some new combination of ideas or problem-elements, and we suggested that humans regularly find these solutions by rearranging problem-elements. It is thus Köhler's stress on a favorable configuration or combination of problem-elements, the prime condition leading to *insight,* which seems more informative than the circularly defined concept of insight itself.

Other Gestalt and Cognitive Approaches: Fixation, Erroneous Assumptions, Direction, and Higher-Order Codes

The difficulty with studying problem solving in animals is that a very large portion of their behavior is determined by basic drives, instincts, and even physiological structure—factors of less importance when considering the human thinker. Descriptions of animal problem solving always deemphasize the main influences on human problem solving described in Chapter 2: habit and conformity.

Unlike Köhler, who in fact stated that unsuccessful problem-solving responses (the direct routes) are biologically determined, later Gestalt and cognitive psychologists focused almost entirely upon the restricting effects of habit upon flexibility. Katona (1940), Duncker (1945), Luchins (1942), Maier (1930, 1970), Wertheimer (1945), and Sheerer (1963) all said little about heredity, but much about the interfering role of past experiences.

Consider Luchins' (1942) mental-set experiments with water-jar problems. The S's task was to obtain a required volume of water, given specific empty jars for measurement. Following a demonstration problem, a group of experimental Ss were given the seven problems shown in Table 3–1. After two or three problems, the average college student detected the pattern $B - A - 2C$ and quickly applied it to the remain-

TABLE 3–1 · Luchins' Water-Jar Problems

Problem	Capacity of Empty Jars			Required Capacity
	A	B	C	
1	21	127	3	100
2	14	163	25	99
3	18	43	10	5
4	9	42	6	21
5	20	59	4	31
6	23	49	3	20
7	15	39	3	18

ing tasks. What's wrong? Problems 6 and 7 were critical test problems that the student could have solved simply by $A - C$ (Problem 6) or $A + C$ (Problem 7). Control Ss who skipped Problems 1 through 5 always found the direct solutions (Table 3–2). Even a habit or mental set established in minutes in the psychological laboratory can blind a person to simpler solutions. Think of the rigidity-producing effects of habits, traditions, and expectations acquired over a lifetime!

About forty years ago, Maier (1933) summarized the following "Hints on How to Reason" intended to combat the rigidity-producing effects of habit: "Do not be a creature of habit and stay in a rut. Keep your mind open for new mean-

TABLE 3–2 · Results of Luchins' Water-Jar Experiment: Performance on Critical Test Problems

Group	No. of Subjects	Percent Indirect Solutions	Percent Direct Solutions	Percent Other Solutions or Complete Failures
Control	57	0	100	0
Experimental	79	81	17	2

ings. The solution pattern appears suddenly. Keep your mind open for new combinations." As we will see in Part III, such attitudinal suggestions still comprise the core of any course in creative problem solving.

Apart from the negative influence of habit, Maier (1930, 1970) names *direction* as the single most important positive problem-solving variable. In his early demonstration, college students were given C-clamps, boards, string, and other paraphernalia, and were asked to devise two pendulums. Students given the direction or hint, "Observe how easy the solution would be if you could only hang the pendulums from two nails in the ceiling," were more likely to find Maier's solution (a T-shaped affair, whose vertical stem was comprised of two boards clamped together, pressing a third board against the ceiling, with a string dangling from either end of the ceiling board) than students without this proper direction.

A later experiment (Maier, 1945) showed how problem cues can provide direction. The pendulum problem was simplified to form a hatrack problem, whose solution was two boards vertically clamped between floor and ceiling, the C-clamp itself serving as the hatrack. When four supports, each identical to the sought-after hatrack, were visible, 72 percent of the Ss were able to construct their hatrack. Without the direction, only 24 percent of the Ss succeeded. I do not suppose we should be too surprised to discover that helpful hints and cues—or direction—facilitate problem solving.

Returning to *habit,* in his 1963 *Scientific American* article, the late Gestalt psychologist Martin Sheerer described the damaging effects of fixating upon misleading problem approaches, perhaps by making incorrect assumptions about a problem. For example, the nine dots in Figure 3–1 must be connected by four straight lines drawn without lifting the pencil from the paper. The problem may be solved only by extending the lines beyond the dots, which most people either do not think of or assume they cannot do. Similarly,

Figure 3–1 · Problem used by Sheerer (1963) to demonstrate effects of incorrect assumptions. The subject's task is to connect the nine dots with four straight lines, without lifting the pencil from the paper.

constructing four equilateral triangles from six matches cannot be accomplished so long as one innocently assumes that the triangles must lie in one plane. The solution—in the form of a pyramid with a triangle-shaped base—requires recentering or shifting one's approach.

Also causing fixation, the individual may fail to detect an object's suitability for a problem because the needed object is hidden in a conventional context. In another of Sheerer's experiments, Ss were asked to place two rings on a peg without coming closer than six feet from the rings and peg. Given two short sticks, Ss easily used a string dangling alone from a nail on the wall to tie the sticks together, creating one stick long enough to perform the task. However, if the string were hanging a No Smoking sign, mirror, or calendar, thus embedded in a meaningful conventional context, over half of his Ss failed to see the string as an available solution-instrument—even though the string was in prominent view and almost everyone decided aloud that he needed just such a piece of string.

Turning now to contemporary cognitive theory, one individual often identified as a cognitive psychologist is Jerome Bruner. We have neither space nor need to review at length Bruner's brilliant and prolific writings on cognitive processes, even though much pertains closely to human problem solving. We shall, however, select those which seem most relevant to this section.

A Study of Thinking (Bruner, Goodnow, and Austin, 1956) is a classic work in the popular research-area of human concept learning. Concept learning is most often conceived as categorizing, a process permitting us to simplify an otherwise overwhelmingly complex world. For example, once we acquire the concept-categories of *ball, rabbit,* or *poison ivy,* or even such abstract concepts as *up* and *truth,* we usually are able to respond appropriately to new examples of these concepts without further learning.

To study human concept learning, Bruner, Goodnow, and Austin presented *S*s with a nine-by-nine array of geometric-patterned stimulus-cards that differed from each other in *shape, color, number of patterns* on the card, and *number of borders* on the card. The college *S*'s problem was to find the stimulus characteristics defining membership in the concept-class. For example, all cards which are *green* in color and *circular* in shape might be members of the concept. The characteristics of number of patterns and number of borders would have nothing to do with this categorization. The *S* could choose any cards in any sequence, and the experimenter would indicate whether or not each chosen card was a member of the sought-after concept-class. Eventually, *S* would learn that green circles were members and all other patterns were nonmembers of the concept.

While the *S*s studied each card and its category, Bruner studied the deliberate *strategies* used by his *S*s to guide their selection of cards to test for category-membership. For example, a *conservative focusing* strategy allows *S* efficiently to eliminate "noisy" or irrelevant stimulus-characteristics one at a time, thereby locating the correct combination of concept-attributes with no wasted card-selections. Bruner's strategies often are taken by critics of learning theory as evidence that the human organism is an active, purposive, and strategic organizer in solving his problems.

In an article closely following the publication of *A Study of Thinking,* Bruner (1957) very carefully described the significance of man's ability to go beyond the information given. Forming an equivalence-class (or conceptual category, or generic code) is one elementary yet common means of generalizing beyond the concrete objects and events. For example, given that an unknown object is a member of the equivalence-class *fruit,* we already know a great deal about the "strange" product: that is, we can go beyond the given information by identifying the new object as a member of a *coded* equivalence-class.

Another form of going beyond sense-data stems from attaching probabilities to relationships among events. Two of Bruner's examples, P*YC*OL*GY or Dwight D. ———, easily allow filling in the gaps based upon learned, probabilistic expectations.

Models and theories represent still another type of higher-order coding that permits accurate predictions and abstractions beyond the information given. The Pythagorean theorem, for instance, describes interrelationships among the three sides and the area of an infinite number of right triangles. The Periodic Table of Elements, a model, allows a near-infinite number of reliable predictions concerning properties of chemical compounds.

The implications for human problem solving of going beyond the information given are implicit and direct, to say the least. Problem solving should become easier whenever an unfamiliar problem can be identified as a member of a class of problems whose solution-strategy is known (for instance,

Cognitive and Gestalt Views

finding the area of a new right triangle). Further, if the problem can be related to a particular theory or model, then these higher-order codes should help clarify and solve the problem.

Incidentally, an important implication for teaching transferable problem-solving skills is that we should emphasize what is generic about a given problem, in order that related problems may be handled more deftly than was the original. That is, we should not emphasize "this solution for this problem," but "this *type* of solution for this *type* of problem" (Bruner, 1957).

Summary and Conclusions

Köhler's classic observations of insight in apes directly or indirectly influenced virtually all later problem-solving research and theory. While his pivotal concept of insight seems poorly defined, Köhler's reference to a good spatial configuration (for instance, seeing the banana and the stick together) as the determiner of insight, is helpful in explaining an important condition for problem solving. If an individual's solution activity or the structure of the problem environment exposes him to component solution ideas, his problem solving (as idea-combining) will be facilitated. Instead of perceptual fields and forces, Gestalt and cognitive psychologists Luchins, Sheerer, and Maier—each with his own demonstrations—emphasized the negative role of habit (taking the form of set, fixation, and rigidity) and the positive effects of direction (hints and cues) in problem solving. Finally, Bruner's conclusions pertaining to "going beyond the information given" directly implied that problem solving may be facilitated by transferring familiar higher-order principles (generic categories, theories, and models) to unfamiliar problem situations.

◄ CHAPTER 4 ►

PROBLEM SOLVING TO THE *S-R* PSYCHOLOGIST: TRANSFER AND TRIAL-AND-ERROR

THERE IS a compelling simplicity in the observation that in problem solving as in other learning the correct response (by virtue of being correct) is rewarded and hence strengthened. Such elegance has led numerous psychologists of human learning to analyze problem solving in the simplifying language of *S-R* psychology. We find, therefore, learning-based problem solving theories that emphasize trial-and-error behavior, habit-family hierarchies, operantly conditioned responses, chains of associations, and response transfer. Typically, while a given theorist ostensibly will focus on just one or two of these interrelated *S-R* concepts and mechanisms, he normally assumes the operation of the others.

Trial-and-Error Problem Solving

Any description of trial-and-error behavior in problem solving must begin with E. L. Thorndike (1898, 1911). Thorndike's landmark research was intended to show the virtual identity of instrumental conditioning (based upon the reinforcement of correct responses) and trial-and-error forms of problem solving (also based upon the reinforcement of correct responses). In his research, Thorndike placed a hungry cat in a jail-like problem box (or puzzle box) that allowed the animal to see and smell an outside dish of food. In different experiments, the animal could escape by turning a door button, pressing a floor lever, or pulling a string. As with Köhler's apes, the cat tested the direct routes first: it tried squeezing through any opening, biting the bars, and thrusting paws and claws through any hole. The animal also clawed and bit objects within the box. Eventually, and apparently by accident, it tripped the device that opened the door. Over about twenty-four trials, the performance measure of seconds-to-solution showed a typical learning-curve: the gradual stamping-in of reinforced responses and the stamping out of nonsuccessful responses. The only apparent difference between individual cats was in their energy level: vigorous, fast-moving animals would more quickly trip the mechanism in the early trials, but not remember the key motion. On the other hand, the slow-movers seemed more attentive to what they did, speeding up solution-times on later trials.

Influenced by the British comparative psychologist, C. Lloyd Morgan, who felt we should never attribute high-level thinking to an animal if the behavior can be explained by a lower-level process, Thorndike concluded that his cats solved (or learned) the problem by mechanical trial-and-error; they were incapable of reasoning and insight, which would have shown an abrupt, not gradual, improvement in performance.[1] He did not, however, claim that humans solve problems only in the mindless trial-and-error fashion of cats.

There are, however, psychologists who strongly and reasonably affirm the importance of trial-and-error in human thinking and problem solving. Woodworth and Schlosberg (1954, p. 818) observed, "Given a genuine problem, there must be exploratory activity, more or less in amount, higher or lower in intellectual level." They further quoted Alexander Bain as noting in the mid-nineteenth century that "In all difficult operations for purposes or ends, the rule of trial-and-error is the grand and final resort."

Donald T. Campbell's (1960) paper, "Blind Variation and Selective Retention in Creative Thought as in Other Knowledge Processes," presents a trial-and-error interpretation of human thought that is rather extreme in the sense that he ultimately reduces all knowledge, wisdom, and inductive processes to blind trial-and-error. His main points, slightly paraphrased, were as follows: (1) a blind-variation-and-selective-retention process is fundamental to all increases in knowledge; (2) the many processes (for example, knowledge of the environment) that shortcut a more complete blind-variation-and-selective-retention process were themselves achieved originally by blind variation and selective retention.

To Campbell, the same "blind-variation . . . etc." underlies and interrelates knowledge gains, problem-solving behavior, and natural selection in evolution. To illustrate, Campbell asserted that the trial-and-error escape behavior of Thorndike's cats actually was not totally "blind" or random. Instead, the animal's squeezing and clawing through holes had either been prelearned by earlier trial-and-error, or had been inherited as a product of (blind) mutation and natural selection. To Campbell, even scientific breakthroughs result from blind explorations: the environment rewards the successful forays, representing discovered wisdom, but does not reinforce the failures. Campbell lists three conditions for blind-variation-and-selective-retention that seem appropriate to natural selection, trial-and-error learning, human problem solving, and perhaps to other forms of knowledge-acquisition: (1) a mechanism for introducing variation; (2) a consistent selection (reinforcement) process; and (3) a mechanism for preserving and reproducing the selected variations. For example, the natural exploratory and manipulative activity of Köhler's chimp Sultan would provide the necessary problem variation; successfully retrieving the banana would selectively reinforce the correct behavior pattern; and Sultan's memory would allow him to reproduce the performance at a later time.

Implicit versus Overt Trial-and-Error in Laboratory Problem-Solving Tasks

In an earlier review of laboratory research in human problem solving (Davis, 1966), I classified tasks according to the type of behavior elicited, overt trial-and-error or implicit problem-solving activity. Looking back on the details of that two-part distinction, the entire issue may be summarized in one sentence: *the problem-solver will think if he can, he will manipulate if he must.*

In many laboratory tasks, the problem-solver simply cannot predict the outcomes of the various response-alternatives, and he necessarily must use the feedback provided by overt trial-and-error responding and by implicit solution-acthinker can remember or predict the outcome or function of each of the response-alternatives, he may proceed by implicitly testing and rejecting these responses and response-combinations (that is, by covert trial-and-error behavior).

One of a series of experiments (Davis, 1967, Experiment III) demonstrated the continuity between problem solving by overt trial-and-error responding and by implicit solution-activity. A switch-light problem-solving apparatus required S to achieve a particular pattern of lights in a matrix by operating switches on his response panel. Figure 4-1, for example, shows a twelve-light, ten-switch problem. Two of the lights, positioned analogously to the move of a chess knight, were red. The S's task was to manipulate his switches until only these two red lights remained on. Each switch controlled two lights, which always were directly adjacent, horizontally or vertically, to each other in the matrix. The solution to the problem in Figure 4−1 involved three switch presses: Switch 6 and Switch 10 each turned on one red light plus an intervening white light, then Switch 3 turned off the two white lights. All other switches turned on lights that eventually had to be turned off in order to solve the problem.

Two groups worked on these problems, one without and the other with pretraining. The Overt Problem-Solving

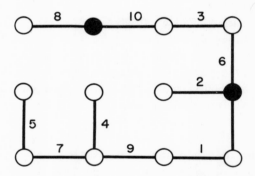

Figure 4−1 · Example of a 12-light, 10-switch problem (Davis, 1967).

Group, without pretraining, invariably approached the task in a fast, overt trial-and-error fashion—since no other strategy could possibly solve the problem. The implicit or Covert Problem-Solving Group, however, was first taught which switches controlled which lights. As expected, errors were few and overt switch presses were slow—the Ss could associate outcomes to the response-alternatives and did not need to behave in a fast, overt trial-and-error fashion. Much like Tolman's 1959 vicarious trial-and-error pointing by rats at the choice point of a T-maze, many Ss in the pretrained Covert Group would grasp a switch without pressing it, then look at the light matrix in an apparent implicit trial-and-error manner.

At present, however, it does not seem important whether the hidden problem-solving activity is called implicit trial-and-error, insight, reasoning, or what have you. It does seem worthwhile, however, to recognize and to be able to predict that S will engage in implicit activity or overt trial-and-error behavior simply depending upon whether he can or cannot associate outcomes to the available response-alternatives.

This simplifying two-part distinction points to one dimension of commonality running through a large number of laboratory problem-solving tasks. Considering just the problems solved by overt trial-and-error, we find a great many classification problems, including the mountain of concept problems, which require the S to sort a large number of stimuli into two or more categories. The concept identification tasks used by Bruner, Goodnow, and Austin (1956), Bourne (1966), Hunt (1962), and many others require Ss to categorize geometric-patterned stimuli on the basis of some experimenter-defined combination of stimulus characteristics. For example, if the concept (or solution) is *red squares*, S is rewarded for attending to color (redness) and shape (squareness) only, ignoring such other attributes as size, number of figures, texture of the figure, number of borders on the card, horizontal or vertical spatial position, etc. Whether we view S as being passively conditioned to the correct *red square* cues and adapted to the irrelevant characteristics (Bourne and Restle, 1959) or whether, since he undoubtedly brings all of his intellectual machinery with him to the laboratory, we say that S formulates strategies (Bruner, Goodnow, and Austin, 1956), tests hypotheses (Bower and Trabasso, 1963), or learns rules (Haygood and Bourne, 1965), the structure of the task invariably demands overt trial-and-error testing—correct responses, hypotheses, strategies, and rules being selectively reinforced.

Probability-learning tasks also must be approached by trial-and-error. In one probability-learning study, Stevenson and Weir (1963) used a three-choice problem in which one response-knob paid off 33 percent of the times it was pressed; responses to the other two knobs were never rewarded. The problem was "solved" when, after considerable trial-and-error knob pushing, S adopted the *maximizing* strategy, always choosing the 33 percent knob. In a hide-and-seek version of probability-learning, Stevenson and Odom (1964) hid toys so that a child would be reinforced on 75 percent of his choices of one box, 25 percent of his choices of another box, and 0 percent of his choices of a third box. On alternative trials, the S hid the toys and the experimenter chose boxes either randomly or according to a fixed 75:25:0 schedule. The S could solve the first "seeking" part of the task by maximizing: always choosing the experimenter's 75 percent reward box. In the second "hiding" part of the problem, he could learn to maximize payoff, at least in the fixed condition, by always hiding toys in the never-chosen box.

In addition to classification problems, probability-learning tasks, and the switch-light problems, a number of other laboratory tasks also require overt trial-and-error approaches. Donahoe (1960) compared two types of feedback in a task which required the S to guess a predetermined point on a seven-by-seven grid. The Ss given one feedback source of "nearer," "farther," or "same" after each guess solved faster than Ss given longitude and latitude information separately, although this latter feedback condition logically would have allowed faster solutions. With a three-part apparatus, Kendler and Kendler (1961) taught children to acquire marbles by operating one of two handles in Device A. A brass cube was dispensed by operating the correct handle of Device B. With Device C, the marble, but not the cube, would earn an M & M candy. When all three devices were then available, Ss inferentially acquired the correct subgoal (marble), using it to produce the major candy goal.

A trial-and-error maze-learning task (Erickson, 1962)

seemed to show the higher-order coding described by Bruner (1957; see Chapter 3). Erickson required two groups of Ss to learn a triangular, spatial-temporal walking maze. "Perception" learners were not blindfolded and were said to learn the maze by spatial and environmental cues. The blindfolded "abstraction" learners were forced to "abstract temporal and directional relations out of the maze environment." While there were no differences between the two groups in learning the original maze, the blindfolded group performed consistently better on two transfer tasks, a rotation of the walking maze and a nine-inch pushbutton maze, both of which required the same sequence of moves ("responses") as the original maze.

Most laboratory tasks evoke implicit, unobservable problem-solving activity. Since we do understand how trial-and-error problem solving works, and since problem solving seems invariably to demand some exploratory idea-combining, it may be helpful to classify these problems as implicit trial-and-error tasks, emphasizing commonality rather than variety. Just briefly, in this group we find our anagram problems, which quite clearly require implicit trial-and-error testing of letter rearrangements (unless paper and pencil or letter blocks are provided). We also find implicit trial-and-error, perhaps less prominently, in Maier's pendulum, two-string, and hatrack problems, Luchins's water-jar problems, a gold-dust version of the water-jar task (Restle and J. H. Davis, 1962), Duncker's (1945) candle problem, and the ball-transfer or pea-transfer problem (Raaheim, 1963; Saugstad and Raaheim, 1960).

Habit-Hierarchies in Problem Solving

The actual responses emitted in overt or implicit trial-and-error problem solving are rarely completely random. Campbell, for example, claimed that the early responses by

Thorndike's cats were determined by prior learning or by heredity. Most learning-oriented psychologists, and many others, would agree that the Hullian habit-family hierarchy mechanism is an attractive and convenient device for explaining the nature and origin of these early problem-solving responses. Furthermore, the habit-hierarchy construct presents a useful definition of a problem and explains the order in which S tests the solution responses, the degree of originality of these responses, and the difficulty level of the problem itself.

In simplified form, we may say that problem stimuli (objects and instructions) elicit a hierarchy of solution responses. The strength or position of each response in the hierarchy is directly related to S's reinforcement history; that is, we assume that the strongest, most dominant responses were more frequently rewarded in similar circumstances than were the weaker responses. The order in which S "thinks of" or tries various responses is directly related to their position in the hierarchy—strongest, least original responses being tested first.

A problem is thus defined as a stimulus situation for which the strongest, first-emitted response is incorrect. (If the first response were the appropriate one, we would simply have an example of stimulus-response behavior, a nonproblem according to the classification scheme of Chapter 2.) A difficult problem is one whose correct response initially is very low in the hierarchy. Conversely, an easy problem is one whose solution initially is more dominant. If the strongest but incorrect response is very, very strong, we have Duncker's (1945) *fixation* behavior, the repetitive perseveration of incorrect behavior that blocks effective problem solving.

The sequential testing and rejecting of response-alternatives may be said to rearrange the S's original hierarchy of response-alternatives by strengthening the initially weak, correct alternative. With the typical laboratory one-shot problem, whose solution you either know or you don't (for instance, anagrams, the pendulum problem, and most others), the low-dominance correct response will jump immediately to the top of the hierarchy. With slower problem solving, as in the case of Thorndike's cats or with Bourne's difficult concept-problems (Bourne and Restle, 1959), the low-dominance correct response gradually increases in strength with successive reinforcements.

Clark Hull (1952) used his powerful habit-family hierarchy mechanism to explain in S-R language the behavior of an organism in Köhler's detour problem. Briefly, with the goal object in view through the barrier, the completely naïve organism will proceed in a straight line toward the object. The vision of the goal object serves as secondary reinforcement maintaining the approach behavior. Physical contact with the barrier discontinues the secondary reinforcement, suddenly causing a radically different set of cutaneous (perhaps painful) stimuli. These latter, unpleasant stimuli are removed by the organism's reflex withdrawal from the barrier, creating a conditioned withdrawal habit to the barrier as a stimulus. Together then, the attraction of the goal-object and the simultaneous repulsion by the barrier lead to locomotor trial-and-error or "exploratory receptor-exposure" acts. Such behavior eventually places the organism far enough to one side of the barrier for an unobstructed view of the goal, evoking uninterrupted locomotion to the object and hence to primary reinforcement.

To Hull, the detour problem involved the limiting case of a habit-hierarchy, one with just two responses: the direct path and the roundabout route. Maltzman's (1955) account of human thinking and problem solving elaborated the Hullian habit-hierarchy to include compound temporal habit-family hierarchies. In these, a given stimulus may elicit hierarchies of hierarchies, a divergent, treelike mechanism.

More simply, we could say that the problem-stimulus elicits several categories of responses. The categories themselves are hierarchically arranged, and the responses within a category also are hierarchically organized. In addition to divergent hierarchies, Maltzman also uses a convergent S-R mechanism, in which several stimuli are assumed to elicit the same response.

Acknowledging that thinking and problem solving include unobservable mental behavior, Maltzman's theory includes the Hullian r_g-s_g mechanism: the environmental stimulus S elicits the implicit response r_g, whose stimulus properties s_g elicit the overt response R—and there will be divergent and convergent hierarchies at every step of the S-r_g-s_g-R formula.

According to Maltzman (1955), "Thinking in general, and problem solving in particular, thus may involve the selection of habit family hierarchies as well as the selection of specific response sequences within a hierarchy." As always, an individual habit-hierarchy is restructured according to reinforcement contingencies, the weak-but-correct response gaining in strength; but with Maltzman's innovation, entire habit-hierarchies also will be strengthened or weakened relative to other hierarchies within the compound hierarchy. To illustrate, consider solving a series of anagram problems whose solutions are words from the same conceptual category (for instance, all are *flavor* words). Not only would individual flavor words, *chocolate, vanilla,* etc., be reinforced, but through mediated generalization the entire habit-family of *flavor* words would be strengthened. In other problems, instructions, mental sets, or hints may serve to increase the dominance of the appropriate hierarchy of solutions.

With the intelligent human problem-solver, the fastest way to increase the strength of his weak-but-correct response is to tell him the solution. Unfortunately, this shortcutting does not leave much problem-solving behavior to be studied, and so researchers have adopted more subtle means of increasing the dominance of the correct solution. Cofer and his colleagues (Cofer, 1957), for example, sought to prime the solution to Maier's two-string problem with appropriate verbal pretraining. In this task, two strings are suspended from the ceiling and S is asked to tie them together. Since they are too far apart for him to grasp one and then walk to the other, he must (with the "pendulum" solution) attach a weight to one string, give it a swing, then grab the other string and catch the pendulum on the return. One group of Cofer's Ss learned eight lists of words, one of which was *rope, swing, pendulum, time, clock*. Compared with a control group, the experimental males (but not females) did indeed produce more pendulum solutions after this verbal pretraining since, according to Cofer, the critical verbal responses were more available.

A number of experiments with anagrams have demonstrated that S's hierarchy of letter-combinations is related to characteristics of the English language, for example, frequency of usage of the solution word and frequency of occurrence in the language of two-letter and three-letter combinations (Mayzner and Tresselt, 1958; Ronning, 1965). In a recent experiment of our own (Davis and Manske, 1968), we thought we could exert more direct control over S's hierarchy of potential solutions. The strategy simply was to have each S memorize a list of words which subsequently would be used in anagram problems. Since S knew that the solution was a single word from the ten-word list, we assumed that he would systematically recall the "hierarchy" of words in their learned serial order until a word was found that exactly matched the anagram letters. We thus predicted that anagram problem-solving time would be directly related to the ordinal position of the solution word in the serial list: words at the beginning of the list should be more quickly available as anagram-solutions than words at the end of the serial list.

There was another possible prediction, however, which turned out to be the correct one. One of the most reliable features of serial list learning is the U-shaped *serial-position curve;* almost always, the first and last words in a list are learned first, and the items just past the middle of the list are the most difficult to learn. It was therefore conceivable that the anagram solution times could be related to the different response strengths shown in the serial-position curve. This curvilinear prediction was clearly more accurate. In Figure 4–2, the anagram solution times plotted over the ordinal po-

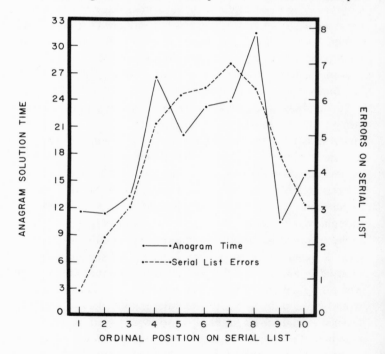

Figure 4–2 · Mean anagram solution times (in seconds) and mean errors to criterion on the serial list as a function of the ordinal position of the solution word on the serial-learning list (Davis and Manske, 1968).

sition of the solution-word in the serial list showed the same inverted-U function as did the serial-list measure of errors to criterion: words learned faster in the serial list (the first and last items) also were solved faster as anagram problems.

In order to clarify the Ss' approach to the anagram task, we simply asked several Ss how they solved the problems—did they actively and systematically recall the earlier-learned serial list, matching each word with the anagram letters, as we previously supposed? Without fail, the Ss reported that they did not bother to systematically recall the list unless they could not solve the problem otherwise. As one S put it, the solution words usually "just popped in." We concluded that there was a clear hierarchical difference in response-availability, related more to the *strength* than to the *order* of prior learning.

Operant Analyses of Problem Solving

A recent operant analysis by Skinner reads as follows, "Problem Solving is concerned with the relations [contingencies] which prevail among three terms: a stimulus, a response, and a reinforcing consequence" (Skinner, 1966, p. 226). He thus reiterated the traditional components of simple operant or instrumental conditioning. Problems, he further maintained, arise when the S-R reinforcement contingencies are complex: for example, competing responses (perhaps emotional) may weaken the correct response or else may destroy the power of the reinforcer; the contingencies may demand a complex chain of responses, which assumes conditioned reinforcers to strengthen S-R links in the chain; or there simply may be no response available which satisfies a given set of contingencies. With these complexities in mind, and with his behavior-control bias showing, Skinner suggested that cats in a Thorndikian problem box could be adapted until their emotional responses disappear. Further, door-opening stimuli could be systematically converted to conditioned reinforcers, which then may serve to shape the cats' movements. Results: both the problem and *S*s' trial-and-error solution behavior would be eliminated.

As we noted in the introductory paragraphs of Chapter 1, if we view problem solving very broadly we may say that the human organism solves problems constantly in the sense of making decisions about immediate courses of action. Skinner interestingly observed that, apart from the slower reinforcement-shaped behavior, the culture provides numerous guides for such frequent problem solving: maxims, proverbs, governmental laws, religious canons, scientific laws, rules of grammar and other rules, policies, heuristics, and algorithms —all serve as discriminative stimuli constructed to evoke correct, reinforcement-producing decisions.

Skinner (1966, p. 256) noted, "An operant analysis [is] often mistakenly identified with stimulus-response theories." How true. Even Arthur Staats, writing in the same volume (Staats, 1966), appeared to use *operant* and *S-R* interchangeably. And we will see that Staats certainly meets Skinner's requirement for a "real" operant analysis of going ". . . beyond input and output to more subtle and more complex environmental arrangements" (Skinner, 1966, p. 256).

To Staats (1968; see also Staats, 1966; Staats and Staats, 1963), problem solving involves the functioning of previously acquired *S-R* repertoires. In his explanation of original verbal behavior and problem solving, these were largely verbal repertoires. Armed with the concepts of word-associations, response-hierarchies, and especially complex environmental stimulus control, Staats first addressed himself to the frequent criticism that *S-R* psychology cannot explain the production of novel sentences. His solution was fairly straightforward: given the existence of two unrelated *S-R* relationships, each established by previous reinforcement, the contiguous occurrence of the two stimuli will elicit a novel response-combination. To use Staats's illustration, if a child has been reinforced for saying "man" in the presence of "man stimulus objects," such stimuli come to control the vocal response "man." Similarly, through past reinforcement, "running" stimuli such as a running dog, running boy, etc., come to elicit the word "running." And although the child never before emitted the response, the occurrence of complex stimuli in the form of a *man running* likely would elicit the novel verbal utterance, "man running."

Adapting this (operant?) analysis of novel sentences to problem solving was conceptually quite easy for Staats. Given that a person has a repertoire of verbal response sequences, each controlled by specific stimulus situations, a complex (problem) situation will tend simultaneously to elicit all of them as a creative problem solution. As with the sentence analysis, it is the configuration that is original, not the component responses.

To more fully account for complex cognitive behavior, Staats, resembling Maltzman, further assumed that convergent and divergent hierarchies of responses, formed by past conditioning, are found at each step of learned verbal sequences. *S-R* chaining also is assumed: simple stimuli produce response sequences which, themselves serving as stimuli, combine with the complex (problem) stimuli to elicit the next response sequence, and so on. Again, each step is assumed to involve hierarchies of response-alternatives.

In addition to "explaining" in operant conditioning language the generation of original verbal behavior and problem solving, Staats (1968) uniquely described such complex thinking and reasoning as *creative scientific behavior*. Consider his example of a creative research chemist. Environmental stimuli—say, the Periodic Table of Elements—produce previously learned complex verbal sequences which describe the systematic nature of the Table. These verbal sequences, again due to past learning, elicit further, more abstract, verbal responses describing the nature of as yet undiscovered elements. And these abstract verbal sequences may lead the scientist to search for, and perhaps find, the undiscovered elements.

Chains of Associations in Problem Solving

To the S-R psychologist, chains of associations (response chains, S-R chains, verbal response chains, and Hull's r_g-s_g mechanism) are devices used to explain the steplike sequence of behaviors leading to the final goal or solution. As we have seen, Skinner (1966) spoke of response chains in problem solving, each link of which must be strengthened by conditioned reinforcers. Staats (1968) argued that natural language word associations and other verbal response sequences (chains) may lead to problem solutions. Cofer (1957) demonstrated the importance of chained verbal units in solving the two-string problem. Hull (1952) assumed that the sight of the goal object in the detour problem elicited an implicit response (r_g) whose stimulus properties (s_g) provided secondary reinforcement for maintaining the approach behavior.

The Kendler and Kendler (1962) model, or pretheoretical model as the authors prefer, centers almost entirely upon chaining. Briefly, two characteristics of problem solving are represented: (1) behavior is a continuous process consisting of chained S-R associations ("horizontal" processes), and (2) at any one time, behavior consists of several ongoing independent habits ("vertical" processes). In the experimental laboratory, horizontal processes may be found in the Skinnerian shaping of long habit chains, in the integration of independent habits by Köhler's chimps, and in studies requiring children to combine separate experiences in order to

Problem Solving to the *S-R* Psychologist 57

solve inference problems (for example, Kendler and Kendler, 1961). The operation of vertical processes is most easily seen in Pavlovian conditioning. The bell stimulus, which elicits an investigatory what-is-it response, occurs parallel (vertical) to the food stimulus which elicits the salivation. These initially independent *S-R* habits eventually interact in such a way that the stimulus (bell) from one *S-R* unit comes to elicit the response (salivation) from the other.

To the Kendlers, problem solving basically follows the pattern of Pavlovian conditioning, except that verbal mediating responses and stimuli are inserted into the *S-R* sequences. An implicit verbal cue from one horizontal chain is said to become associated, or vertically connected, with the correct overt response from another. Serendipity, finding one thing while seeking another, was considered a good example of how the connection of a stimulus and a response from independent chains may produce problem solutions.

More generally, we might speculate that creative problem solving by metaphorical thinking, as described in Chapter 9, conceivably may resemble the Kendlers' vertical-horizontal model. For example, a bat or porpoise may navigate by bouncing its voice off unseen objects. This information, viewed as a horizontal idea-chain, might suggest a solution for the parallel (vertical) stimulus problem of inventing a "navigational aid" for blind people.

Transfer in Problem Solving

It is a tautological truth that "appropriate" past learning will improve problem solving and inappropriate experiences will not facilitate (in fact, may impair) finding a solution. This observation led Schulz (1960) to recast problem solving in terms of transfer, which may be positive (facilitative) or negative (inhibitory). For example, Schulz viewed fixation behavior, as demonstrated by Luchins's water-jar problems, as a matter of negative transfer: practice with the first set of problems interferes with finding a simpler solution to later problems.

In the psychological literature, transfer of training almost always is studied with verbal materials, using the standard paired-associates (P-A) procedure. From this literature has evolved a very large number of principles, laws, and variables pertaining to transfer. The beauty of viewing problem solving as transfer of training, as does Schulz, is that these principles, laws, and variables identified in P-A research may shed light upon factors (for instance, stimulus and response similarity) influencing problem solving.

Conclusions

Despite their different emphases, all of the specific S-R learning theories described in this chapter assume that problem solving and thinking obey the same powerful laws of conditioning as do simpler forms of learned behavior. Certainly, there is no doubt that much human problem solving is tremendously clarified and objectified when prior learning and present reinforcement contingencies are examined—it is no accident that the basic reasoning behind modern learning theory has survived 2,500 years since Aristotle's laws of association.

There are, however, at least two important shortcomings of all of the S-R views of problem solving summarized in this chapter. The first difficulty is that in gaining simplicity, the S-R theorist sacrifices completeness. By dissecting human thinking and problem solving into the theoretical language of simple conditioned responses, the theorist must ignore too much beautifully conscious and deliberate mental behavior. Does any conditioning analysis thoroughly describe the skills, strategies, thoughts, and emotions (for example, happiness) of the human as he writes a poem, discovers a physical law, or even plans dinner? To even attempt a complete picture of such activities, one must use a phenomenological language which accepts the complexity of the conscious, thinking, and feeling human being.

The second problem is that we find precious little of prescriptive value in the learning-based problem-solving literature. Always describing and explaining, the monolithic *S-R* machinery does not specify how to become a more skillful, more imaginative problem-solver. The suggestion that we reward original responses is not enough, despite laboratory demonstrations with humans (Maltzman, 1960) and creative porpoises (Pryor, Haag, and O'Reilly, 1969).

NOTE

1. We will shortly explain that whenever a problem-solving organism cannot predict the outcomes (or correctness) of the solution-alternatives, he necessarily must adopt a trial-and-error testing strategy. Köhler (1925) criticized Thorndike's research and conclusions on much the same basis, arguing that the necessary solution components were hidden: the cat could not see nor understand the mechanical door-opening contrivance. Köhler further noted that the required responses were unnatural, since cats normally do not escape by turning or pressing a lever. He concluded that Thorndike's cats could solve this task only by sheer, blind chance. If the solution components were within the animals' view and comprehension, and the response within the animals' normal repertoire, the cats may have shown the sudden insight (the adoption of a roundabout route) displayed by chimps.

◄ CHAPTER 5 ►

COMPUTER PROBLEM-SOLVING MODELS: A TOOL TO MATCH THE TASK

IN CONTEMPORARY PSYCHOLOGY and other disciplines, the word *model* profitably may be taken to mean analogy. In any model or analogy, there exists an isomorphism, a point-for-point correspondence between the model and the phenomena being modeled. A model train, to take a simple illustration, will correspond point-for-point with the real train. The quality or worth of the model, be it a train or a theoretical model, depends upon the "goodness of fit": the more accurately the details of the model correspond to and predict the details of the real object or process, the better the model. A model therefore cannot be judged "true" or "false" in any meaningful sense; it may be evaluated only in relation to how well it fits or parallels the real phenomenon.

A number of computer scientists interested in psychological problems have noted that in a great many respects a computer may serve as a model of complex human behavior. Hunt (1970), especially, describes in detail his structure and process model of complex human behavior, derived from and empirically supported by computer-simulation research.

There are, certainly, many point-for-point analogical correspondences between computer information processing, on

Computer Problem-Solving Models 61

one hand, and the sensory input, "thinking," and motor responses of the human, on the other. For example, a computer may receive information from any of several "sensory" channels, along with instructions for performing operations on this input. The machine has a "memory" or information-storage capability, and it can "search" through stored lists of information, inserting and "forgetting" (erasing) as required. Every computer "makes decisions" by comparing specific information against preprogrammed decision criteria; it can even "learn" in the sense of cumulatively storing new information, instructions, and decisions, using this newly acquired input to make "more informed" decisions. At the output end, the computer usually spews forth printed data at its familiar blitzkrieg rate; but it also may "talk," draw a picture or graph, or otherwise display its alphanumeric knowledge.

Problem-Solving Programs:
Algorithmic and Heuristic

The generic term *artificial intelligence* encompasses all computer programs that play games (cards, chess, checkers, and others), retrieve facts, recognize figural patterns, do mathematics (algebra, geometry, integral calculus), learn concepts and paired-associates lists, make decisions based upon various combinations of rewards probabilities, and complete letter or number sequences (e.g. ABACADA . . . ?). It is important, however, to distinguish between algorithmic and heuristic programs. The algorithmic program will produce as output humanlike solutions without necessarily proceeding through a humanlike sequence of operations. By brute computing force, the algorithmic program is designed to generate and examine all possible solution-combinations in some predetermined order.[1] Eventually, and inevitably, it locates the correct solution-alternative, terminating the search. Most checker-playing and chess-playing programs are written to play well, but not necessarily imitate the perceptive, shortcutting behaviors of the human player (Hunt, 1968).

The heuristic program, on the other hand, proceeds through its searching and decision making roughly as the human information processor might. The forte of the heuristic program is its ability to drastically reduce the amount of mechanical trial-and-error searching by selectively pursuing only goal paths that appear promising.[2] Unlike the powerful algorithmic programs, the heuristic program possesses the definitely human ability to fail, usually by prematurely eliminating the correct path to the desired conclusion. Naturally, the heuristic simulation programs are more helpful than the algorithmic in explaining the nature of complex human problem solving.

Developing the Simulation Program

In an oversimplified and highly condensed form, the computer model builder proceeds about as follows. He first proposes a relatively precise sequence of problem-solving operations or processes that he thinks would minimally permit the human thinker to solve a particular problem. Normally, the programmer identifies these individual human information processes and their organization by requiring human Ss to "think aloud" while solving actual problems. A set of computer subprograms or subroutines is prepared, each of which can execute as often as needed a specialized process corresponding to each of the postulated human processes. For example, various subroutines may transform or classify input data, perform many kinds of calculations, make comparisons and decisions, use decision results to guide memory searches, compare calculations against solution specifications, etc.

The strategic ordering and use of the various subprograms is guided by the programmer's calling or executive program, which also is based upon the human Ss' think-aloud protocols—plus a large dash of programmer intuition.

To be judged successful, the program usually must meet two criteria: first, it must solve the problem; second, its printed-out sequence of problem-solving operations should reasonably duplicate the processes of the typical human performing the same task. If the programmer wishes to test further the predictive worth of his program, he can modify parts of it and then compare the machine performance with that of the human working under analogous problem modifications.

The beauty and true scientific contribution of computer-simulation research is that such work demands absolutely that the programmer, a theoretician, state precisely and completely a sequentially organized set of problem-solving processes that will successfully complete the task. If the information fed in, or the set of operations on that information, is in any way incomplete, the computer will let the programmer know about it. A mindless follower of instructions, the machine will either print back nonsense, get lost in a feedback loop, or refuse to run.

The Program As a Theory

According to Newell, Shaw, and Simon (1958), a program that successfully models (or simulates, duplicates, imitates) the tricks, devices, and strategies of the human problem-solver may be viewed as a theory that explains that human behavior. In the remainder of this chapter, we will examine some representative computer-simulation programs. Each program, as a theory, seeks to explain the observable human problem-solving behavior by specifying the needed information and an organized set of information-processing operations.

The Logic Theorist

The landmark program that introduced computer-simulation of complex behavior to psychology was the Logic Theorist (LT) of Newell, Shaw, and Simon (1958). Their goal was to program a Rand Johniac digital computer to prove the set of theorems in formal logic derived by Alfred North Whitehead and Bertrand Russell in *Principia Mathematica* (1925). Using IPL-V, a computer language for operating on lists of symbols, the computer specialists first prepared a set of methods or general schemes for discovering proofs. For example, substitution was always tried first: C is sought, and B implies C, then substituting the proof of B for the proof of C solved the theorem. Or, one step further, if B implies C and A implies B, then substituting A proved the theorem. If substitution proved futile, the program called, in order, the next three inference-rules of detachment, forward chaining, and backward chaining.[3]

In addition to the inference rules, the computer's storage banks were fed the axioms of Whitehead and Russell. Further, the program specified that as each theorem was proved, it was stored along with the axioms for future use, just as Whitehead and Russell had done. In a sense, the machine was programmed to learn or improve its problem-solving skill by progressively accumulating the experiences it would need for tackling increasingly complex problems.

Finally, the first fifty-two theorems in Chapter 2 of *Principia Mathematica* were read into the computer in the exact order in which they appeared in that volume. Its success? Overall, LT proved thirty-eight of the fifty-two theorems, about 73 percent. Also, the printed-out sequence of operations was judged by Newell, Shaw, and Simon to adequately mirror the think-aloud protocols of human Ss working on the same problems.[4]

In a series of followup runs, Newel, Shaw, and Simon found that—just as with any human problem-solver—success depended closely upon the availability of essential information: if some intermediate proofs were skipped, problem solving was difficult or impossible; if all preceding proofs were available, the task usually was completed.

The authors also observed other characteristics of human problem solving in the computer's performance. For example, the particular sequence of methods used by LT (substitution, detachment, forward chaining, then backward chaining) were said to constitute a directional or method "set," analogous to the human mental set. Newell, Shaw, and Simon also commented upon the ever-present trial-and-error versus insight issue. Specifically, the amount of mechanical trial-and-error searching was effectively reduced by heuristically using goal information to work backwards, preferentially testing only axioms and theorems that were in some way similar to the desired proof. Such selective testing seemed to resemble a meaningful trial-and-error strategy, yet the final solution displayed the characteristic suddenness of insightful problem solving.

In sum, LT could find proofs for the Whitehead and Russell logic theorems—it possessed and specified at least a minimally sufficient organization of information processes. As with the human problem-solver, LT's performance critically depended upon the presentation sequence of the cumulatively stored problems; and LT displayed humanlike insight in the sense of employing strategies or heuristics for reducing trial-and-error search behavior.

The General Problem Solver

While the performance of the Logic Theorist was indeed impressive, it could solve only those problems for which it was specifically designed, discovering proofs for theorems in symbolic logic. Revising LT and its variants, Newell and Simon (1961) produced the more powerful General Problem Solver (GPS), a program capable of playing chess, proving logic theorems and trigonometry identities, solving word puzzles, and even writing computer programs—including some of its own subroutines. GPS thus may be applied to a large variety of problems, so long as these problems can be recast to include an original or starting state, S_o, a goal state, S_g, and an admissible set of operations.

The strategy of GPS was to compare S_o and S_g, obtaining a set of differences. The differences serve as an heuristic to guide the search for a sequence of operations that would remove these differences. The solution, in fact, is that sequence of operations that "maps" or moves the original state S_o to the goal state S_g. Incorrect sequences of operations lead to blocked paths, causing the computer to back up to an earlier choice-point. To illustrate, in proving formal logic problems or trigonometry identities, there exist numerous formulae and axioms known to be true that, in their many potential combinations, collectively comprise S_o. Inference-rules, also a given, serve to *operate* on the formulae, locating an ordered subset of these formulae that solves the problem: that is, removes the differences between S_o and the desired S_g. In chess problems, the initial board position defines S_o, S_g is the set of positions checkmating the opponent, and the operators are defined by the legal moves of the game that (literally) "move" S_o to S_g.

The GPS also has the humanlike capability to momentarily set aside the main problem while clearing up a subproblem. As Hunt (1970, p. 24) metaphorically described the GPS context-switching capability, ". . . if I wish to go from Seattle to London, I must take an airplane, and to do this I must get to the airport." As another human problem-solving characteristic, GPS is most efficient if the operators are listed in order of potential usefulness.

Problem: In its increased power and generality, compared to the Logic Theorist, does the General Problem Solver better simulate the information processes of the human thinker? Hunt (1970), referring to GPS as a (nonsimulating) algorithm, conceded that data are not available to evaluate GPS as a psychological theory: that is, as an accurate description of the nature and organization of human mental operations. It is possible, for example, for GPS to accept and solve problems in ways that would be unfamiliar and very difficult forms of thinking for the human.

The Perceiver

In discussing the values of computer-simulation work, Simon and Newell (1956) noted that the model builder is forced to think very seriously about the ways humans think and solve problems. One computer model, Simon and Barenfeld's (1969) Perceiver, illustrates very clearly how simulation work forces or stimulates the theoretician to postulate in detail some elusive mental behaviors. The reader likely will detect that in this section more emphasis will be upon identifying actual human thinking activities and less upon programmed operators.

The Perceiver model sought specifically to clarify human perceptual-organizational processes in chess playing. Simon and Barenfeld observed that during the first fifteen seconds or so that an experienced player views a new chess position, he does not immediately search for various profitable moves.

Rather, he is carefully noting the essential properties of the position—gathering information that will help him anticipate the consequences of various moves.

The authors proposed a rather specific explanation, or theory, of the cognitive processes underlying the player's eye movements: at each point of eye-fixation on a new chess position, the player acquires information about the location of pieces at or near the fixation point. He also learns about pieces in peripheral vision (within approximately 7° of arc) that are related in an attack, defend, block, or shield manner to the piece at the point of fixation. Simon and Barenfeld further assumed that (1) information being gathered concerns meaningful relations between pieces (usually two) or between one piece and a square, and relates to defending and attacking; and (2) when eyes are fixated at a particular point, attention may shift to a meaningfully related neighboring piece, and back again, without a change in actual eye fixation. Perceiver was the translation of these assumptions into a computer program, using elementary information processes borrowed from another chess-playing program, Mater (Baylor, 1965; Baylor and Simon, 1966).

After "viewing" a chess position, Perceiver printed out its simulated sequence of eye movements. Nicely supporting the above theoretical assumptions, Perceiver's eye movements were very similar to human chess-player eye movements recorded by Tichomirov and Poznyanskaya (1966). One interesting exception: while both human and computer focused on the ten most critical pieces, the human (but never the machine) occasionally fixated on an unoccupied square.

Having thus specified, and empirically supported, the cognitive processes underlying a chess-player's eye movements, Simon and Barenfeld proceeded to explain short-term memory for chess positions. Interestingly, they did no further programming. The authors merely combined concepts from Perceiver with ". . . another component of the information processing theory of cognition . . . the EPAM (Elementary Perceiver and Memorizer) theory of rote learning (Gregg and Simon, 1967)," a model that incorporates Miller's (1956) chunking or encoding hypothesis of memory.

First of all, Simon and Barenfeld noted that after a player views a chess position for a few seconds, his ability to accurately recall that position depends critically upon two factors: the *meaningfulness* of the position and the *skill* of the player. A chess master will easily reproduce a meaningful position almost without error, while a weaker player can accurately replace only about six pieces. But when confronted with a random board, the master can correctly replace no more pieces than the novice (DeGroot, 1965). The EPAM-based explanation (or theory) ran about as follows: with a *meaningful* board, a master will recode or "chunk" the large number of pieces and positions into a smaller number of familiar constellations—putting the entire board within the human short-term memory capacity. Less skilled players, almost by definition, do not see these constellations and therefore try to memorize a large number of individual pieces and positions. The novice's memory burden is thus considerably greater. With random boards, the absence of familiar constellations reduces the master's encoding ability to the level of the weaker player, about six pieces.

Computer simulation work does indeed stimulate careful thinking about the ways humans perceive, learn, and solve problems—with or, as in this case, without actually writing and running a program.

Other Simulation Programs

In addition to the Logic Theorist, General Problem Solver, and Perceiver, numerous other programs have been written to simulate various types of human problem-solving behaviors. All of these are heuristic in the sense of using humanlike rules of thumb to reduce the amount of trial-and-error searching for the solution. Two mathematics programs resemble LT and GPS in using powerful heuristics for guiding the selection of subproblems. The Artificial Geometer (Gerlernter, 1959) proves theorems in plane geometry, while the Saint program (Slagle, 1963) produces solutions to problems in integral calculus. Slagle and Dixon (1969), Green (1969a, 1969b), and J. A. Robinson (1965, 1969) have described general programming strategies for proving theorems, playing games, and solving other problems.

Several programs simulate human concept-learning. As we have seen, in experimental psychology learning a concept is equivalent to learning to *classify* different objects into two or more groups (members and nonmembers of the concept class) on the basis of certain object-characteristics. Hunt 1962), Hunt and Hovland (1961), Baker (1964, 1965, 1967, 1968), and E. S. Johnson (1964) all have constructed programs to simulate the strategies that humans use to deduce the characteristics defining the concept-class.

Number and letter sequence problems (for example, ABACADA . . . ?) used in intelligence testing for years, have also been programmed. Simon and Kotovsky (1963) prepared a program that, much like the human thinker, searched for cycles of progressively longer length. Once the cycle or cycles were discovered, additional members of the sequence could be generated.

Another common intelligence-test item is the analogy, and again programs exist that try to solve analogy problems using operations which approximate those of the human. Based upon verbal-association networks, Reitman's *Argus* (Reitman, 1965; Reitman, Grove, and Shoup, 1964) could solve word analogies (for example, China is to Communism as the U.S.A. is to ———) by searching for a preprogrammed relationship that would link Word *A* with Word *B*,

Computer Problem-Solving Models

and that also would link Word C with one of the D solution-alternatives. Evans (1968) wrote a program for solving geometry analogies. The program establishes one or more relationships (or *mappings*) between Figures A and B. These relationships are then applied to Figure C in an effort to produce one of the D-alternatives (that is, to "map" Figure C into Figure D).

Apart from the Perceiver program for simulating chess-player eye movements, Newell and his partners (Newell, Shaw, and Simon, 1963; Newell and Simon, 1965) partly completed a highly sophisticated program for playing chess. Other programs by Simon (Baylor and Simon, 1966; Simon and Simon, 1962) solve classical chess problems. According to Hunt (1968), however, programs for playing *complete* chess games may not exist, despite many press releases (for example, pitting American against Soviet programs) to the contrary.

The Creative Computer

One issue with computer-simulation of problem solving is whether or not such computer behavior might be called creative in any meaningful sense. Acknowledging the vagueness of the term *creativity,* Newell, Shaw, and Simon (1962) decided, "Creative activity appears . . . simply to be a special class of problem solving activity characterized by novelty, unconventionality, persistence, and difficulty in problem formulation." Furthermore, they observed that their problem-solving programs carry on activities ". . . not far from what is usually regarded as 'creative' " (Newell, Shaw, and Simon, 1962, p. 68). For example, they argued that if it was creative for Whitehead and Russell to write *Principia Mathematica*—one of the most significant intellectual products of this century—then possibly their Logic Theorist program was behaving creatively in reinventing large portions of Chapter 2 of the volume, often rediscovering the very same proofs found originally by Whitehead and Russell. With one theorem, in fact, LT found a shorter, more elegant proof than did Whitehead and Russell.

One naturally resists attributing such human capabilities as creativity, intelligence, or genius to a machine, which obviously behaves only as programmed, but the computer performance does meet most definitions of creative behavior. Or does it?

Conclusions: Cognitive-Gestalt, *S-R,* and Computer-Simulation Views of Problem Solving

Any attempt to somehow compare or evaluate the relative contributions of cognitive-Gestalt, *S-R,* or computer-simulation theories must consider the historical background and goals of each viewpoint. The learning theorist, for the last half-century, has sought to understand complex behavior by reducing it to pure simplified elements: namely, principles and processes of conditioning. For example, in his psychological laboratory the *S-R* theorist might study the classical conditioning of human eyelids or the rote learning of nonsense syllables. Higher mental capabilities that might lead the human subject to voluntarily shut his eye, thus avoiding the air puff to the cornea, or to use mnemonic tricks to memorize the word list, were considered to contaminate the research—voluntary responses in eyelid conditioning would not be scored; mnemonic devices in rote memorizing would be ignored.[5] As we noted at the conclusion of Chapter 4, the learning theorist has gained scientific simplicity—and the concepts of trial-and-error learning, habit-hierarchies, operant conditioning, transfer, and *S-R* chains indeed are comparatively simple—but his oversimplified descriptions of thinking and problem solving suffer badly in completeness.

Computer Problem-Solving Models

The cognitive-Gestalt viewpoints described in Chapter 3 and the computer-simulation theories of the present chapter both try to make sense of human thinking and problem solving in a more thorough and realistic fashion than is the case with S-R approaches. The theoretical languages of the cognitive-Gestalt and the computer-simulation theorist are alike in the basic freedom to speculate on the purposive thoughts and strategies of the problem-solver. Insight, fixation, direction, problem assumptions, higher-order codes—all are concepts describing conscious mental activities of the same order of cognitive complexity as the computer theorist's classifying, recalling, comparing, deciding, and using heuristic shortcuts.

Beyond their common-sense data-language level, however, the verbal theories of the cognitive psychologists obviously are of a different nature than the program-theories of Hunt, Newell, Simon, and other computer specialists. Indeed, the capability of the computer to act as an analogue of the human thinker opens new worlds of detailed completeness unavailable to the laboratory psychologist, learning or Gestalt. As we have noted, a successful program will provide a detailed, unambiguous description of a problem-solving sequence that parallels, point by point, the strategic way a human might solve the same problem.

In the chapters that follow, dealing with the training in creative problem-solving skills of students, engineers, and executives, the concern is with the whole, intact organism consciously learning complex skills and attitudes. Therefore, as with the cognitive-Gestalt and computer-simulation theories, our language will be a phenomenological, common-sense one. One could, I believe, analyze the effects of these training programs and courses in the language of conditioned reflexes, mediated generalization, transfer, and so on. However, such an effort would amount only to a reductionistic academic exercise unlikely to aid our understanding of the emergent, cognitive dynamics of real-world creating and problem solving.

NOTES

1. Newell, Shaw, and Simon (1957) claim that the algorithmic strategy is about as sensible as placing typewriters before monkeys and waiting until they produce all the books in the British Museum.
2. The main strategy for evaluating "promising" goal paths is to work backwards from the goal—thus assuring that every sequence the machine considers can, in fact, lead to that goal.
3. See Newell, Shaw, and Simon (1958) for a detailed explanation of each inference rule.
4. In his recent review of computer-simulation studies, Hunt (1968) concluded that the comparisons described by Newell and his associates ". . . simply do not meet reasonable criteria for supporting experimental evidence." It was not clear, noted Hunt, just how often the human protocols matched the machine performance.
5. In the last decade, at least, mnemonic devices themselves have been the focus of much research. See, for example, Paivio (1969) or Rowrer and Ammon (1971).

◀ PART III ▶

The Focus on Training: Industry and Education

WITH THIS SECTION we begin describing some highlights of our second body of problem-solving literature, that dealing with the training of creative and problem-solving skills. For continuity with the psychological literature of Part II, our first topic concerns *inquiry* activity. A fairly well-defined research area in education, the inquiry approach treats problem solving as sensing a puzzling event that is then clarified, researched, and finally resolved. Like basic psychological research, many inquiry studies focus on clarifying task and subject variables influencing human inquiry; others aim directly at teaching skills of systematic inquiring.

Professional creative problem-solving methods always teach attitudes conducive to creative behavior and concrete techniques for the conscious production of new idea-combinations. The best-known deliberate creative problem-solving procedure, brainstorming, is sketched in Chapter 7. The key brainstorming principle is deferred judgment, an attitudinal concept that creates the free, receptive atmosphere necessary for truly novel thinking.

While brainstorming largely provides favorable psychological conditions for flexible problem solving, there are techniques that specify more directly how new idea-combinations may be "forced." The *attribute listing, morphological synthesis,* and *idea checklist* procedures (Chapter 8), along with the metaphor-based *synectics* and *bionics* (Chapter 9), are all techniques for stimulating new idea-combinations that may be used in either group or individual problem solving. It is important to note that virtually all of these methods were derived from introspective reports of creative individuals. The methods may be viewed then, as "unconscious" strategies made knowable.

It is this clarification of idea-finding that is so importantly different from problem solving as it is conceived and studied in the psychological laboratory. Rarely has any theory or research in psychological problem solving specified how distantly related solution-components become combined into a final creative solution. Further, many idea-finding methods —for example, those described in Chapters 7 through 9— are sufficiently well-defined to allow them to be taught to others.

There are skeptics, informed and ignorant, who feel that genuine creative ability is determined by genetics or God, and therefore cannot be taught. Meanwhile, back at the corporation, engineers and individuals at all levels of management claim that they are learning to be more flexible in their thinking and more receptive to change and innovation; they learn individual and group methods for producing imaginative problem solutions, and when the training is over they feel more creative and are generally more self-confident.

The Focus on Training

When measures of creative performance are available, such as the number or quality of innovative ideas adopted or the number of patents applied for, the worthwhile effects of training are usually further confirmed. Sylvania Electric, to quote an extremely favorable evaluation of one training program, reported, "Doubled profit; 2,100 new products; beat competition on two new products; increased patent applications five-fold; saved $22 million" (Edwards, 1968).

In the schools, effects of programs intended to teach creative problem solving are more difficult to evaluate. Ideally, such training would improve long-term problem solving in the student's personal and educational life, and would ultimately lead to a more creatively productive and satisfying career. Obviously, evidence on these points is unavailable to the program developer with average patience. Of necessity, researchers resort to tests of creativity and other problem-solving tests (see Appendixes A and B), attitude surveys, classroom observations, and reports of teachers and students in order to evaluate training effects. Despite difficult measurement problems, however, educational researchers have devised methods and materials for encouraging some definable problem-solving skills. In addition to the inquiry training of Chapter 6, Chapter 10 summarizes our Wisconsin Project. Three Wisconsin programs, two teaching general creative thinking and problem solving and one teaching creative writing, have drawn concepts from the professional problem-solving literature and from many other sources. Chapter 11 summarizes still other strategies and materials for fostering problem solving and creative thinking by changing attitudes in a more flexible direction, by teaching problem-solving steps, and by exercising creative thinking abilities.

◄ CHAPTER 6 ►

INQUIRY ACTIVITY: STUDY OF THE COMPLEAT PROBLEM-SOLVING EXPERIENCE

IN STUDYING PROBLEM SOLVING, psychologists and computer scientists typically present the subject with a carefully defined task, such as anagrams, a chess problem, or an unusual-uses task, and then record one or more selected problem-solving responses. While this controlled laboratory strategy is well suited for its purpose of identifying basic variables and processes, the act of meeting and solving a problem in the real world is rarely so neatly structured. Rather, the encounter with a problem seems more to resemble a sequence of activities described by Dewey (1938) and more recently by Shulman (1965, 1967; Shulman, Loupe, and Piper, 1968) as (1) *problem sensing,* in which a person initially detects, to his discomfort, that some kind of problem or incongruity exists; (2) *problem formulating,* wherein the person subjectively defines a particular problem and develops his own anticipated form of solution; (3) *searching,* in which the individual questions, hypothesizes, gathers information, and occasionally backtracks; and (4) *problem resolving,* the final phase in which the person becomes satisfied that he has solved the problem or "found out why," thus removing the disequilibrium.

Educational researchers have adopted the word *inquiry* to identify research literature pertaining to the overall problem-solving experience described by the Dewey-Shulman stages. The inquiry model presents a refreshing and realistic approach to the self-initiated and self-directed form of natural problem solving, which most educators feel should be encouraged in the classroom. Indeed, while the *S-R*, Gestalt, and computer models largely describe problem-solving variables and processes, the inquiry concept describes both real-world problem solving, or inquiring, and attempts to prescribe classroom experiences to foster skills important for inquiry.

Research in inquiry activity reflects this dual function of describing and prescribing—or studying and training—the phenomenon of human inquiry. On one hand, researchers such as Shulman, Allender (1969), Butts (1965), and Koos (1969) have sought mainly to understand inquiry abilities, strategies, and behavior patterns, and to devise tests to measure these abilities and behaviors. On the other hand, Suchman (1960, 1963, 1966a), while naturally intrigued with the nature and measurement of inquiry activities, has primarily directed his efforts toward prescribing teacher and student activities for strengthening inquiry skills.

Examining and Measuring Inquiry Abilities and Behavior Patterns

Shulman (1965, 1967; Shulman, Loupe, and Piper, 1968) for several years has studied the strategies and abilities involved in inquiry activity. His research will illustrate both the complexity of natural human inquiry and the wealth of information to be extracted from well-done inquiry research.

Shulman's own first task was to create a problem environment meeting several criteria: The situation must be intrinsi-

Inquiry Activity

cally motivating, it must contain a large number of potential problems,[1] and it must provide for the observation and recording of several dependent measures. The *teacher's in-basket* seemed to satisfy these requirements very well.[2] The subject, an ego-involved female teacher trainee, is placed at a desk whose in-basket is well-loaded with potential problems. She is asked to play the role of a teacher who suddenly takes a sixth-grade class in the middle of a semester. No students actually are present (because of a convenient "holiday") and the trainee therefore ". . . may begin where she likes and do as she pleases . . . with no time limit" (Shulman,1967).

In addition to the contents of the in-basket (phone messages, notes from faculty and administrators, personality test and sociogram information), the teacher has available students' report cards, aptitude and achievement scores, attendance records, and miscellaneous anecdotal reports. Further, by phoning a "school secretary" or a "principal" (confederates who play fixed roles) the teacher may obtain past grades, medical records, and family information such as socioeconomic level. Still another information source was named the *reference ego,* a voice which the teacher could consult by phone as she might her own memory. The reference ego knew all about such things as local administrative policy and teacher gossip.

Virtually hundreds of potential problems were embedded in the in-basket and in the other information sources. To illustrate, one memo simply asked if the teacher had any referrals for the school psychologist; and if so, would the teacher please describe the problem and her hunches about its source. This memo could lead the teacher to discover, for example, one epileptic student whose family cannot afford seizure-preventing medication. The materials on this epileptic student alone would provide several subproblems which the teacher might sense and resolve in order to deal with the larger problem. Another in-basket memo reports that one student cannot pay for his books until his father finds a job. A problem such as this may not be sensed as problematic, it might be sensed but deferred, or the teacher may try to resolve it. Clearly, the in-basket technique was structured to elicit complex, subject-defined, and subject-determined inquiry activities.

With the aid of tape recorders, observers behind one-way mirrors, think-aloud protocols, and postsession interviews, Shulman obtained a rather substantial amount of quantitative and qualitative information. On the quantified side, he recorded such performance scores as (1) *problem sensitivity,* the number of embedded, potential problems to which the teacher responded; (2) *materials attended,* the number of different pieces of material used by the teacher in the inquiry period; (3) *information sources,* the types or categories of information used by the teacher; (4) *time,* the number of teacher-determined minutes spent in the inquiry session; and (5) *conceptual tempo,* a finer analysis of latency patterns including, for example, changes in the rate of problem sensing or information processing.

One may also make qualitative observations and decisions in such research. To begin with, judges simply estimated the teacher's "competence," an index of *problem-resolution ability.* Shulman further identified two main approaches to the in-basket problem: the *one-by-one* serial method, in which the teacher works on problems in the order she senses them, and the *surveying* approach, where the teacher first tries to attain an overall grasp of the situation before attacking any one problem. Partly independent of her approach, the teacher may adopt one of three sequence-strategies. While the *consecutive seeker* works out one problem to completion before tackling another, the *simultaneous seeker* efficiently groups problems on the basis of required search activities, then works on several problems at once. Finally, the *branched seeker* begins with a single problem, but then branches out to a new and more compelling task, only to branch again and again as her interests and attention direct her. At any time, the branched seeker will have several problems at various stages of completion.

Overall, Shulman concluded that traditional laboratory problem-solving research indeed is pitifully inadequate in regard to studying real human behavior in real problematic situations. The world just does not present carefully defined and controlled problem-solving episodes. Shulman's main findings pertaining to human inquiry dealt with both the characteristics of the subject and the task. For example, the best inquirers—that is, those who were sensitive to problems, used a wide variety of information sources, and were rated as highly competent in inquiry—were found to be politically liberal, associatively fluent, cognitively complex, sound interpreters of written passages, reflective, and non-anxious. Further, effective inquirers spent more time in inquiry activity, during which they engaged more in problem-solving activity, than in nonproblem-oriented survey activity. Finally, Shulman observed that among his sample of teacher trainees, an individual's grade-point average was not related to his predisposition for successful inquiry behavior.

Allender (1969) described an inquiry strategy resembling Shulman's teacher's in-basket in the form of materials entitled *I Am the Mayor* (Allender and Allender, 1965; Allender, 1968). Like Shulman, Allender also wished to study the abilities, behaviors, and motivations underlying human inquiry.

Closely following the four-part model of inquiry—problem sensitivity, problem formulation, search behavior, and resolution—the *Mayor* materials contain four sections: (1) *the mayor's work,* 10 documents (letters, phone messages, a newspaper, and various reports) that could cross the desk of a mayor of a city of about 12,000; (2) *the mayor's questions,* about 80 pages of multiple-choice formulations of potential problems; (3) *the mayor's file,* containing about 250 pages of calendars, laws, maps, records, city plans, and correspondence; and (4) *the mayor's decisions,* about 30 pages of possible resolutions.

Allender's "Mayors" were sixth-grade, seventh-grade, and eighth-grade students who ". . . were told to do what they wanted to do because [after all] they were the mayor" (Allender; 1969, p. 548). For dependent measures, Allender recorded scores corresponding to the inquiry components: *problem sensitivity* was quantified as the number of problems sensed (number of question pages asked for); *problem formulation* was scored as the number of questions asked (number of times the mayor's file was used); *search behavior* was a count of the number of file pages viewed.

One of Allender's major conclusions was simply that independent inquiry activity is a very normal learning process— at least when conditions allow it to occur. Motivated by intrinsic task interest, rather than by such extrinsic reinforcers as money, points, grades, or even feedback on solution correctness, all fifty-one students sensed problems, asked questions, requested information, and reached decisions until the secretary-experimenter explained that there was no more work for the mayor to do. Allender's study would strongly support Suchman's (1960) claim that students motivated to "find out why" respond better to this than to any other classroom incentive.

Butts (1965; Butts and Jones, 1966) described a clever instrument for measuring inquiry skills, the *Tab Inventory of Science Progress* (TISP). In a slight reformulation of Shulman's four components, the TISP assumes that when a child inquires into a puzzling situation, he searches, processes data, discovers, verifies, and then transfers learned concepts to new situations. In five distinct phases, the TISP (1) presents a physics problem (on film) and (2) asks the student to describe in his own words, for example, why the problem

Inquiry Activity 85

apparatus did not work. In Phase 3 the student is allowed to select prepared yes-no questions, which may be *irrelevant, inadequate, redundant,* or *relevant,* and then (4) the child is asked again to explain the physics problem. Finally, (5) the student is asked to explain a similar problem, providing a measure of concept transfer. A comparison of the S's solutions in Phase 2 (before his question-asking) with his solutions in Phase 4 (after the questioning), along with a sequential record of his actual questions, provides measures of inquiry problem-solving skill.

Another measure of inquiry habits and skills is Afsar and Koos's (1970) twenty-item *Behavioral Checklist for Science Students* (BCL). Each student in a high school science class is asked to rate other students in his class according to such inquiry-related characteristics as showing high enthusiasm for science problems, conducting projects with speed and accuracy, asking questions that show understanding and imagination, consulting many information sources, and proposing inventive problem-solving methods.

Koos and her colleagues at the Mid-Continent Regional Laboratory also constructed a novel test of biological inquiry skills entitled *Explorations in Biology* (EIB) (Burmester, Garth, Koos, and Stothart, 1970a, 1970b).[3] The EIB booklets allow the student to lead himself through an actual bit of biological research. The initial inquiry-evoking, puzzling event is a set of six slides. The first, labeled "Summer, 1957," presents a scene near a field of ripe corn. Leaves on a tree and vegetation around the field are green. A number of adult wild land-based birds are perched in the tree and on the ground. The second and third slides, entitled "Summer, 1959" and "Summer, 1961," are very similar to the first slide, with all birds appearing alive. Beginning with "Summer, 1963," however, a few dead birds appear on the ground. More dead birds are shown in "Summer, 1965," and by "Summer, 1967" all birds but one are dead.

In a unique form of branching program, Booklet A of the workbook-like test allows the student to choose the order of steps he would take, if he were a biologist, to investigate the mysterious increase in bird deaths. For example, he first might choose to (1) read a list of possible explanations of the event, then (2) read potentially useful reports about birds, then (3) read a list of relevant questions about birds, and finally (4) read a list of ways to get new data to explain the event. Regardless of the order in which S selected these steps, each alternative would branch him to a multiple-choice situation. If he began with Step (1), he would choose two out of ten explanations that he might investigate first—again, if he were a biologist. Following (2), he would read and rate the usefulness of four of ten reports about birds. Step (3) would let him pick the "most important" of five questions to ask about the puzzling event. Step (4) asks him to rate the soundness of ten different ways of getting new data about the dying birds. The alternatives in each of these multiple-choice situations are differentially weighted in their contributions to the student's overall biological inquiry ability score.

In Booklet B, *Explorations in Biology* asks the student to select an experimental design that would unambiguously test whether or not Pesticide X could have caused the bird deaths, and to interpret a series of graphs displaying possible research outcomes. Overall, Koos (1969) and her colleagues feel that the EIB test provides a reasonable assessment of important and, it seems, very complex biological inquiry skills.

Training Skills of Scientific Inquiry

Credited with pioneering the recent interest in inquiry training, Suchman (1960, 1963, 1966a, 1966b, 1967, 1968, 1969) employs his inquiry strategy to motivate the learning of physics and science concepts while stimulating the growth

Inquiry Activity

of scientific problem-solving ability. His specific strategy might be illustrated as follows: elementary school children observe a short demonstration, either on film (Suchman, 1960) or live (Suchman, 1966a), of a curious cause-and-effect relationship in science. Curiosity thus aroused, the children essentially perform an imaginary experiment by posing yes-no questions for the teacher. In one report (Suchman, 1960), the children have seen a film showing that a heated metal ball will not pass through a ring, while the same ball when cooled passes through the ring quite easily.

PUPIL: Were the ball and ring at room temperature to begin with?
TEACHER: Yes.
PUPIL: And the ball would go through the ring at first?
TEACHER: Yes.
PUPIL: After the ball was held over the fire it did *not* go through the ring, right?
TEACHER: Yes.
PUPIL: If the ring had been heated instead of the ball, would the results have been the same?
TEACHER: No.
PUPIL: If both had been heated would the ball have gone through then?
TEACHER: That all depends.
PUPIL: If they had both been heated to the same temperature would the ball have gone through?
TEACHER: Yes.
PUPIL: Would the ball be the same size after it was heated as it was before?
TEACHER: No.
PUPIL: Could the same experiment have been done if the ball and ring were made out of some other metal?
TEACHER: Yes.

Naturally, some information-gathering and hypothesis-testing questions are better than others. Suchman (1960) therefore created his *episode-analysis* strategy to coach students in asking pointed questions. First of all, they were encouraged to identify important problem *objects, systems* (of two or more objects), *properties* of objects and systems, *conditions* or states of objects and systems, and *events* (changes in the conditions of objects and systems). According to Suchman, these episode-analysis concepts aid students in overcoming their natural tendency for Gestalts—that is, for perceiving the situation as a complete, nondissectable whole. Suchman further showed students that by strategically altering one *object* or *event* at a time, they could isolate the critical factor determining the puzzling cause-effect relationship. Such simplification would allow the students to formulate rules pertaining, for example, to the scope of the episode and its determining factors.

Suchman (1960) reported that after fifteen weeks of scientific inquiry training, fifth-grade students demonstrated both an increase in their understanding of the course content and a new proficiency in using the scientific method. They tended to make fewer untested assumptions, they better formulated and tested their hypotheses, they more carefully designed their experiments, and they more efficiently discovered causal relationships and regularities.

As a final suggestion for conducting successful inquiry sessions, Suchman recommended that regardless of whether the inquiry period ends in success, failure, or—most likely —expiration of time, a postsession review of strategies and tactics is indispensible both for maintaining morale and for maximizing the training of inquiry behavior.

Conclusions

It seems evident that human inquiry as summarized in this chapter describes fairly well the total problem-solving experience as it most often is encountered in our real world. The

Inquiry Activity

human problem-solver, whether concerned with the selection of his breakfast cereal, with rewriting an awkward paragraph, or with deciding his future career, will sense the difficulty, try to clarify it, explore alternatives, and reach an equilibrium-restoring solution. The interest shown by education-based researchers in clarifying inquiry patterns and strategies, in developing tests of inquiry skills, and in producing strategies and materials for classroom inquiry training, seems appropriately worthwhile.

NOTES

1. It is important to distinguish between the experimenter-defined potential problem and the S's sensed problem.
2. The creation of the *teacher's in-basket* stems from the earlier *in-basket test*, a measure of administrative decision-making developed by Frederickson, Saunder, and Wand (1957).
3. Like the *teacher's in-basket*, *I Am the Mayor*, and the *Tab Inventory of Science Progress, Explorations in Biology* very likely strengthens skills of inquiry during the process of studying or measuring them.

◄ CHAPTER 7 ►

BRAINSTORMING

ANYONE INTERESTED in problem solving or thinking should try group brainstorming at least once—just to watch the birth of surprising and imaginative ideas. Undoubtedly the most popular form of "forced" creativity, brainstorming is used regularly to prod flights of professional imagination, especially by advertisers who must persist in trying to catch the attention of our seriously disinterested public. A flexible teacher also will schedule classroom brainstorming sessions, letting her charges regress, pretend, and fly where they will —habits of nearly every highly creative adult.

Reasons for the very wide acceptance of brainstorming are quite clear: it's intuitively appealing, it's simple, it's fun, it's therapeutic, and it works. The intuitive appeal stems from the single hard-and-fast rule, the *principle of deferred judgment*—the self-evident notion that criticism and at least harsh evaluation will interfere with flexible idea production.[1] We all know other people, usually those higher in rank, who can effectively crunch every potentially constructive idea that positive thinkers might suggest. Aside from the punishing effects of criticism, it also seems evident that one cannot evaluate and think of more ideas at the same instant.

Brainstorming is simple because, again, the only strict rule is the deferment-of-judgment principle. Since no training is necessary beyond a few minutes' clarification of the ground rules, even elementary school children may exercise their creative abilities while perhaps solving some local problems of, for example, traffic safety, building and grounds sanitation, lunchroom disturbances, tardiness, or flat cafeteria food.

As far as entertainment goes, brainstorming often is so much fun that professionals feel guilty about being paid for having such a swell time. We have noted that a creative idea may be viewed as a new, surprising combination of previously unrelated ideas. Humor also is based upon new, surprising juxtapositions of elements. It is no accident, therefore, that ideas produced in brainstorming sessions or creativity-training courses often are humorous.[2] Even though a few ideas will be absurd (except in their capacity to further stimulate new viewpoints), some will be unique, solid, interesting ideas, to which someone in the group inevitably will respond, "I wonder why no one ever thought of that before?"

Can brainstorming be therapeutic? There is considerable frustration among individuals caught in an overstructured bureaucracy. University students as well as workers and managers are rarely asked for their ideas or opinions on how the job, the company, the classroom, or the university ought to be run. There is a basic need to think and create that is almost never satisfied on the job or campus. Even at ostensibly more satisfying executive or administrative levels, the chance to speak up, kick around ideas, receive feedback from others, solve pressing problems, and have fun, can be therapeutic. Barron, in his books *Creativity and Psychological Health* (1963) and *Creativity and Personal Freedom* (1968), explains at length the mental-health value of freedom to create.

And brainstorming works, too, whether the goal is to stimulate imagination and flexibility as a training exercise—

for children or professionals—or to solve some elusive industrial problem. Osborn (1963), Mason (1960), and Clark (1958) describe numerous instances of successful ideas emanating from scheduled brainstorming sessions. While his list is naturally selective, Osborn (1963) reported that a Denver postmaster and eleven of his staff brainstormed the problem, "What can be done to reduce man-hour usage?" Some of the 121 ideas led to a saving of 12,666 man hours in the following nine weeks. A Pittsburgh department store, stuck with some chair-covering material, brainstormed "other uses" for the fabric. The resulting ideas led to advertising which sold the entire stock in a week. A brainstorming group at Heinz Foods spent just one hour on the problem, "How can we help increase the sale of products made at this factory?" They generated more and better ideas than a special committee had produced in ten ordinary conferences. Reynolds Metals brainstormed some "New and more convenient ways to package Aunt Jemima Cornmeal Mix." Some of the resulting ideas were included in a prize-winning and sales-winning package. Sylvania Electric's highly successful *Flash Cube* also was born in a brainstorming session. Finally, the success of the New York advertising agency Batten, Barton, Durstine, & Osborn—Alex Osborn's own organization—probably is due in some part to their use of brainstorming.

Four Ground Rules

Formally, Osborn (1963, p. 156) lists four ground rules, all of which are intimately tied to the deferred-judgment principle—which, of course, is ground rule number one. Since these rules amount to critical conditions for creative problem solving, we will comment briefly on the functional and psychological significance of each:

1. *Criticism is ruled out.* Adverse judgment of ideas must be withheld until later.

Brainstorming

Since the goal of the group is to produce a large number of ideas, members naturally must feel free to do so. In Chapter 1 we described some attitudes contributing to creative behavior. A receptive, constructive attitude will more effectively stimulate problem solving than a fault-finding negative one—the difference is reinforcement versus punishment. In progressive education, one often hears of the creative atmosphere, which is little more (but never less) than an encouraging, receptive attitude by an enlightened teacher. The most striking characteristic of a brainstorming session is its creative atmosphere, stemming from the flexible attitudes of the participants.

2. *Freewheeling is welcomed.* The wilder the idea, the better; it is easier to tame down than to think up.

This second ground rule nicely supplements the deferred-judgment principle, since a deliberate search for wild, imaginative problem solutions requires a receptive, noncritical atmosphere. Encouraging wild ideas recognizes that creative ideas, by definition, are unusual, imaginative problem solutions. The brainstormer—as any other creative thinker—must be consciously set to be imaginative, to try different and unusual strategies, and to view the problem from novel perspectives—in a word, to suggest anything. To illustrate, a thoughtful brainstormer working on a within-factory transportation problem (How do we get things from here to there?) might suggest, "Train cockroaches, there's plenty of the _____ things around!" Such a wild idea realistically might suggest some cockroach-like, under-the-floor conveyance system, or perhaps electronic cockroaches of various sizes that might be programmed to scoot from place to place as directed. In one of my own recent classroom brainstorming exercises, the suggestion that a movie theater could be emptied quickly by collapsible seats that slide patrons out the front of the room was immediately followed by the sounder idea of seats that fold down so patrons could easily walk over them.

Everything we know about mental associative behavior supports Osborn's recommendation for wild freewheeling. Such stimulation stirs one's imagination and at the same time activates the associative capabilities of others. Are obviously wild and absurd ideas wasted? Frankly, yes. It is expected that of the final list of ideas produced in an active group-think session, many of which are wild and impractical as per instructions, about 5 percent will be worth further exploration. But just one or two good ideas may more than justify the "wasted" 95 percent.

3. *Quantity is wanted*. The greater the number of ideas, the greater the likelihood of useful ideas.

This principle restates the charge of the group—to produce as many ideas as possible. Statistically, it is obvious that as the total number of ideas increases, the number of potentially valuable ideas also will increase, given that the proportion of good ideas remains constant.[3] Furthermore, the greater the number of ideas listed, the more they can enter into combination with other ideas, mushrooming the total idea output. Psychologically, it is true that ideas, as with any other associations, become progressively more original when more and more ideas are listed (Manske and Davis, 1968; Parnes, 1961). In the movie theater improvement problem mentioned above, the first ideas were for more comfortable seats and for more leg room. Forty minutes and about eighty ideas later, some suggestions were for an electronic seating chart in front of the theater, automatically indicating with lights which seats still were available; telephone ordering and pneumatic tube delivery of refreshments to your seat; and conveyor-belt seats that pick you up in the lobby and later deposit you on the front sidewalk at the end of the flick.

4. *Combination and improvement are sought*. In addition to contributing ideas of their own, participants should suggest how ideas of others can be turned into better ideas; or

how two or more ideas can be joined into still another idea.

We decided in Chapter 2 that much of the mystery surrounding creative production is reduced if we assume that creative ideas are new combinations of previously unrelated ideas. By recognizing the importance of combining ideas, Rule 4 instructs thinkers to deliberately seek idea-combinations and modifications, thus supplementing the spontaneous ideas for a greater total idea-pool.

The four ground rules are explained to any newcomer in a brainstorming session, along with procedural details. For example, some groups keep a small bell in the center of the table—to ding anyone who criticizes. Some leaders prefer to go around the table, letting each participant speak in turn, but allowing someone with a pressing idea to interrupt. Other leaders let the members respond and interrupt as they see fit. Ideas may be recorded by a stenographer—or better yet, two—or by a tape recorder. In my opinion, the best method is to list ideas on a blackboard where they are available for extension and recombination.

Other Procedures and Recommendations for Brainstorming

While a brainstorming session may be conducted in the most informal and casual fashion, so long as the principle of deferred judgment is followed, Osborn (1963; see also Clark, 1958; Mason, 1960) proposed many experience-based suggestions for the professional who may wish to initiate brainstorming: (1) Group size should be ten to twelve members. The ideal twelve-man panel will consist of a leader, an associate leader, about five regular members, and about five guests. (2) To increase the source and variety of ideas, members should be heterogeneous in training, experience, and sex. In most circumstances, rank must be as equal as possible. It is easy to imagine the stifling effects of a top executive, especially one unacquainted with brainstorming principles, sitting in on a lower-status think-session. (3) Members should be advised of the problem forty-eight hours in advance of the session, allowing them to think about the task and come in with prepared ideas. (4) The session itself should last just thirty to forty-five minutes, time enough to exhaust everyone. (5) During the sessions, the leader may handle the uncomfortable silent periods (if any) by asking the less outspoken members for their ideas, by suggesting some solutions of his own prepared for just such an emergency, or by referring to an idea-stimulating checklist such as Osborn's (1963) "seventy-three idea-spurring questions." (6) About two days after the session, it is profitable to circulate a request for postsession ideas and afterthoughts. A critical presession consideration, incidentally, is whether or not to use a group to solve the problem in the first place. Even Osborn recognizes that, "Despite the many virtues of group brainstorming, individual ideation is usually more usable and can be just as productive" (1963, p. 191).

Variations on a Theme

There are two refinements that may increase brainstorming effectiveness in particular circumstances. The first is *stop-and-go brainstorming* in which short (about ten-minute) periods of unrestrained freewheeling are interrupted by periods of evaluation. The latter serve mainly to keep the group on target by selecting the apparently most profitable directions. The second is the *Phillips 66* technique for use in problem solving by large audiences. After the problem is clearly understood, small groups of six will brainstorm for six minutes, after which one member of each group reports either all or the best ideas to the person directing the show.

Problem Definition

Perhaps the most critical feature of organizing a successful group problem-solving session lies in initially defining the problem. The questions asked will surely determine the answers received. First of all, the problem must be simple, focused on a single target. If a problem contains too many complexities, it should be broken into subproblems, each to be the subject of a problem-solving session. For example, if the original problem were "Think of ways to improve a kitchen stove," the problem might be too complex. Such a statement could be broken into subproblems: for example, improve the surface burners, think of new colors and color patterns, improve the oven insulation, think of original rearrangements of the stove components, and so on. At the same time we are simplifying the problem, however, we also must define the task broadly enough to allow totally new approaches to the task. For example, if a rodent problem were stated as "How can we build a better mousetrap?" we limit our view to one approach: trapping. More broadly stated, the real problem is "How can we get rid of mice?" which opens new avenues beyond just designing new traps.

With our stove-improvement task, consider the subproblem, "How can we improve the surface burners?" As stated, one would be limited to thinking of changes and improvements for good old-fashioned surface burners. We might therefore think of making them square instead of round, making them extra small (to heat one cup of tea) or extra large, adding more gradations to the heat levels, or putting doors over them to add counter space. Now, a broader statement of the real problem might be, "Think of better ways to heat food," which opens up entirely new possibilities. Perhaps surface burners are unnecessary. Electric heating coils certainly would be more efficient if they were hidden in the bottom of the pan, which conveniently could be placed (or plugged in) anywhere on a colorful electricity-conducting surface. If we eliminate electric-coil heat entirely, perhaps infrared, electronic, or other heat sources could replace surface burners.

We could continue this entertaining free-association behavior, but the point is just that the problem must be simple and focused, yet broad enough to free the imagination. Problem definition will determine not only the nature but the imaginativeness of the solutions.

Idea Evaluation

Brainstorming occurs in at least three different settings, each with its own purpose: (1) in the classroom, where an artificial problem may be used to stimulate and exercise imaginations; (2) in industrial or other professional situations, where real and practical problems are attacked; and (3) in research on creative problem solving, in which the scholar-researcher studies environmental or task conditions for successful brainstorming. The evaluation of brainstormed ideas is naturally quite different in each of the three situations. In classroom brainstorming, whether in an elementary school or in a professional creative problem-solving class, there may be no need for a formal evaluation of the ideas at all. Since the purpose of the group is to practice brainstorming, or to practice leadership of a problem-solving group, the listed ideas, while interesting, typically are just scrapped. It is possible, of course, that school children could list worthwhile solutions, for example, to a local traffic or sanitation problem. In this event, they would no doubt enjoy passing on some of the intuitively more sound ideas to the principal or school board.

The evaluation of brainstormed ideas in the business

Brainstorming

world naturally is quite different. Some time after the thinking session, group members (or a separate group) will scrutinize each idea according to such criteria as marketing appeal, workability, production costs, availability of raw materials, acceptability to higher management, possible patent infringements, and so on. Sometimes, a list of criteria may be brainstormed in a separate session. A few (or perhaps none) of the ideas will survive this first sounding to be sent on for further examination, model development, and perhaps implementation. The final criterion of ideas for product development or cost cutting usually is the single *post hoc* one: Did it make or save money?

In creative problem-solving research, the creativeness of ideas produced by a given brainstorming group may be measured in any or all of a variety of ways. Most common are scores for idea fluency, originality, and flexibility. Except for fluency—a simple count of total ideas listed—there is virtually no standardization in scoring, unless one uses as a brainstorming problem an exercise from the published Guilford or Torrance tests, for which scoring directions and norms are provided (see Appendix A). The following, then, are some methods that I and others have devised for evaluating idea-lists. The discussion assumes that the creativeness of one brainstorming group's idea-list is to be compared with the creativeness of several other lists (from other groups that worked on the same problem, but perhaps under different instructional or training conditions).

After omitting from a given list both duplicated ideas and ideas which are judged to be totally unrelated to the problem, (1) idea fluency is scored simply by counting the number of remaining ideas. (2) Originality may be scored several ways, most simply by counting the number of ideas that are completely unique; that is, that do not appear on any other list. Originality also may be scored on the basis of statistical frequency. An idea which occurs just once in the total pool of ideas from all lists receives a higher originality score than an idea occurring twice, three times, etc. Without reference to other idea-lists, judges may rate each idea on an "originality" scale. (3) One may further score for flexibility by counting the number of different categories of ideas produced. (4) "Practicality" (usefulness or feasibility) also may be scored with rating scales.

One may use these four measures to determine still further dimensions of creativeness. We may wish to (5) count the raw number of ideas scoring high on originality, or (6) compute the proportion of original ideas (number original/total number of ideas in a list). (7) One might also be interested in a count of the number of ideas scoring high in both originality and practicality. (For research purposes, ideas that are both original and practical may be considered "good" or "creative" ideas.) (8) The proportion of creative ideas (number creative/total number of ideas in a list) also may be helpful. (9) If the thinking period is timed, one may divide the group's total number of ideas produced (fluency score) by the number of minutes of the session, producing the speed measure of ideas per minute, whose inverse is response latency (minutes per idea). Similarly, one may derive (10) original ideas per minute or (11) creative ideas per minute. A measure devised by Warren (1970) directs the researcher to compare the best one, two, three, four, or five ideas of one group with the best one, two, etc., ideas of each other group. (12) Finally, if the group or individual determines its own stopping point, the number of minutes spent brainstorming provides a rough index of motivation (Warren and Davis, 1969).[4]

On the Negative Side: Qualifications and Limitations

Thus far, our reaction to brainstorming has been mostly glowing. The basic concept is sound, and there are many testimonies of success. Realistically, however, Osborn himself does not guarantee his group brainstorming as a cure-all for every idea-needing institution. Indeed, some brainstorming programs completely fail to provide the creative breakthroughs that had been anticipated. Osborn notes two main reasons for at least some disappointments. First, the organizers and members may not be following Osborn's brainstorming procedures, and second, they may unrealistically expect miracles in the first place (Osborn, 1963, p. 191). All forced creative thinking techniques discussed in this volume, including brainstorming, should supplement, not replace, original individual thinking.

Some controlled research has concluded that brainstorming does more harm than good. Typically, these studies show that the deferred-judgment principle (and by implication, brainstorming) does not aid creative problem solving, or that individuals are more creative when alone than in groups. Weisskopf-Joelson and Eliseo (1961), for example, found that instructions to wildly generate imaginative ideas are less productive than instructions to give good practical ideas. Also, Taylor, Berry, and Block (1958) and Dunnette, Campbell, and Jaastad (1963) decided that despite deferred-judgment instructions, individuals are in fact inhibited in the group sessions because of fears of criticism. Osborn's reply to these studies was simply that brainstorming sessions do produce ". . . more and better ideas than is possible through the usual type of conference in which judicial judgment [*sic!*] jams creative imagination" (Osborn, 1963, p. 168).

Conclusion: Conditions for Creative Problem Solving

Osborn's brainstorming illustrates simple yet sound conditions for problem solving with imagination. Perhaps most pertinent is the creative atmosphere—a matter of attitudes—that will exist if the ground rules are followed. Members may freely entertain farfetched ideas, recognizing that the wild thought might be tamed into a workable solution or might stimulate others to create further ideas. Also, brainstormers deliberately (and often spontaneously) produce combinations and modifications of earlier ideas which, as described in Chapter 2, is how many new ideas are created. The quest for quantity recognizes that later ideas are most often the best ideas. While brainstorming is not guaranteed 100 percent effective, it just might be more creatively productive than the average committee meeting.

NOTES

1. Actually, *brainstorming* may be defined by the presence of the *deferred-judgment* principle even in individual brainstorming. If criticism is scrupulously avoided, you are brainstorming. On the other hand, brainstorming *qua* brainstorming ends with the first criticism or evaluation.
2. Conversely, humorous ideas typically could be considered creative.
3. The proportion of good ideas often does remain constant (Warren and Davis, 1969).
4. For more details of these scoring methods, see Davis and Manske (1968), Warren and Davis (1969), or Davis and Roweton (1968).

◄ CHAPTER 8 ►

STRATEGIES FOR STIMULATING SOLUTIONS: ATTRIBUTE LISTING, MORPHOLOGICAL SYNTHESIS, AND IDEA CHECKLISTS

THREE METHODS for producing novel idea-combinations are attribute listing, morphological synthesis, and using idea checklists. Instruction in the use of these three techniques is included in most professional creative problem-solving courses. The methods may be used to extend intuitive idea finding in individual problem solving or in group sessions such as brainstorming.

Attribute Listing

The man associated with the attribute-listing strategy is the late Robert P. Crawford, formerly Professor of Journalism at the University of Nebraska. In Chapter 1 we credited Professor Crawford with initiating in 1931 the very first structured course in creative thinking. His main book, *Techniques of Creative Thinking* (1954),[1] is a classic in the field of creative education, along with Osborn's (1963) *Applied Imagination* and Gordon's (1961) *Synectics*.

Crawford's attribute listing, which is both an idea-finding technique and a theory of creative behavior, may be summarized in one sentence (Crawford, 1954, p. 96), *"Each time we take a step we do it by changing an attribute or a quality of something, or else by applying that same quality or attribute to some other thing* [italics in original]." To Crawford, then, original invention in any field occurs by the stepwise improving of attributes (or parts, qualities, characteristics) of a given object, or else by transferring attributes from one object or situation to a new object or situation. Objects includes technological, literary, musical, and artistic "objects."

Any item, simple or complex, may be improved by isolating and modifying individual attributes or qualities of that object. As a simple illustration, consider a piece of common classroom chalk, which has the obvious attributes of *size, color, shape,* and *hardness.* One may easily design an innovative line of chalk by modifying each attribute: *Size?* How about super-small, super-large, or double-length? *Color?* In addition to all of those in the rainbow, one might look around for novel gold, silver, brass, or copper colors, perhaps fluorescent yellow, orange, and chartreuse, and maybe even special-purpose striped chalk or chalk that draws a dotted line. As for *shape,* the possible modifications are limitless and could be indefinitely explored with a couple of ounces of putty (perhaps animal shapes would stimulate children to draw).

In designing clothes, attribute listing almost is the *modus operandi,* deliberately or not. Consider the creation of a shirt or blouse: the cuffs, sleeves, collar, "cut," colors and color patterns, material, and closure device are separately considered attributes that may be modified or perhaps removed.

The application of the attribute-modifying strategy obviously is very straightforward. The reader may wish to turn his eyes to any object and begin permuting different *shapes, colors,* and *materials.* The industrial specialist might further see and mentally modify such attributes as function, cost, reliability, serviceability, or availability of raw materials. In a slightly less honorable form, Crawford's attribute modifying resembles closely the standard substitution method, used by appliance designers to effortlessly outdate your refrigerator every year—by systematically altering such attributes as color, size, capacity, cooling potential, shelf and door handle design, or by substituting materials: plastic for metal, glass for plastic, nylon for glass, and so on.

In addition to the strategy of changing attributes, Crawford describes how transferring attributes can lead to creative problem solutions and innovation in general. To use a few of his illustrations, creative architecture might involve transferring some attributes of, for example, Mediterranean architecture to a previously non-Mediterranean setting. A few strains of folk music may provide the foundation for a hit tune. A touch of cartoonism may inspire a sequence of paintings possessing a creative touch of both irony and humor. A museum display of historic ladies' and men's clothing may (in fact, does) suggest innovative fashions for today. The possibilities for transferring attributes are extensive, to say the least, given that the innovator can recognize and develop a good bet when he encounters it.

We will end this section with a few testimonial sentences from Crawford: "Perhaps you do not catch the full import of what we have been doing. . . . Being original is simply reaching over and shifting attributes in what is before you" (Crawford, 1954, p. 52). "The process of creation is so simple and so easy, when one understands it, that even the best of things is usually susceptible of improvement" (p. 96).

Morphological Synthesis

The morphological synthesis procedure, originally developed by astrophysicist Zwicky (1957) in his book *Morphological Astronomy* and then refined by Allen (1962, 1966), amounts to a logical extension of attribute listing, although the two strategies no doubt evolved quite independently.[2] In brief, regardless of the qualitative nature of the problem, one first identifies two or more major dimensions (or attributes) of the problem. To use our shirt-designing problem mentioned above, seven dimensions might be cuffs, sleeves, collar, cut, colors and color patterns, material, and closure device. Second, one lists ideas for each of these dimensions. In the present example, without even trying to be imaginative we could list four cuffs (one-button, two-button, French cuffs, none), two sleeves (regular, puffed), five collars (plain, button down, "mod-long," Nehru, none), three cuts (regular, body-hugging, ruffled), twenty colors and color patterns (take your pick), six materials (cotton, rayon, silk, satin, velour, muslin), and three closure devices (button, zipper, none—pullover style). Finally, one evaluates the huge number of all possible idea-combinations. This exceedingly modest morphological synthesis produces 43,200 possible combinations. Because none of the individual ideas are really wild, many of the 43,200 combinations could be appealing.

If just two or three dimensions are used, one can draw a two-dimensional or three-dimensional diagram, in which each cell represents an idea-combination. In our sixth-grade to eighth-grade training program, *Thinking Creatively: A Guide to Training Imagination* (Davis and Houtman, 1968; see Chapter 10), the cartoon characters in the story line invented new ice cream flavors by listing 56 flavors on one axis and 36 "extra-goodies" (nuts, fruits, candies, vegetables, etc., to go in the ice cream) on the other axis of a two-dimensional "checkerboard." The result was 2,016 ice cream flavors, about half of which were feasible (e.g., honey-flavored with pistachio nuts, coconut-flavored with gum drops), although many were predictably repulsive (e.g., spinach flavored with plums). In another venture, the cartoon characters invented 182 vehicles by listing sources of power (e.g., atomic, gas motor, sails, steam, pedal) along one axis, and body type (e.g., car, bus, train, flying saucer, bicycle) along the other axis. While a pedal-powered flying saucer may never get off the ground, or the drawing board, the atomic train seemed like a good original (to the cartoon characters) idea. One sixth-grade class, participating in a field test of *Thinking Creatively,* used the morphological synthesis or checkerboard method to find theme-writing ideas for their entire class—by listing short-story characters along one axis of the checkerboard and settings along the other.

Idea Checklists

Very generally, whenever we browse through a pet store for new tropical fish, consult a thesaurus to find the right word, read the Yellow Pages to locate the nearest TV repairman, or scrutinize a catalogue, we are using idea checklists. Thus, as the name implies, the checklist strategy simply amounts to examining some kind of "list" that could suggest solutions suitable for a given problem. The "theory," if one is needed, stems from our definition of a creative idea as some new combination of previously unrelated ideas. An idea checklist essentially stimulates nonobvious and nonhabitual idea-combinations.

Checklists can extend our intuitive idea supply, first of all, by directly providing solution possibilities. Browsing through a shop, a catalogue, or a thesaurus would be a direct means of finding problem solutions. On the other hand, checklists may indirectly stimulate the production of new ideas far beyond what is provided in the list itself. It is the latter, imagination-prodding quality of the checklist strategy that has led to its extensive use in courses, books, and programs for training creative problem solving in industry (for example, Mason, 1960; Osborn, 1963; Parnes, 1967; Simberg, 1964), and in some educational training programs (Davis and Houtman, 1968; DiPego, 1970).

CHECKLISTS FOR STIMULATING NEW IDEA-COMBINATIONS

Osborn (1963) devised his "seventy-three idea-spurring questions" to inspire individual or group brainstormers. Consider, as you read through this checklist, how a mousetrap, can-opener, or some other common object might be reshaped by Osborn's hints:

Put to other uses? New ways to use as is? Other uses if modified?

Adapt? What else is like this? What other idea does this suggest? Does past offer parallel? What could I copy? Whom could I emulate?

Modify? New twist? Change meaning, color, motion, sound, odor, form, shape? Other changes?

Magnify? What to add? More time? Greater frequency? Stronger? Higher? Longer? Thicker? Extra value? Plus ingredient? Duplicate? Multiply? Exaggerate?

Minify? What to subtract? Smaller? Condensed? Miniature? Lower? Shorter? Lighter? Omit? Streamline? Split up? Understate?

Substitute? Who else instead? What else instead? Other ingredient? Other material? Other process? Other power? Other place? Other approach? Other tone of voice?

Rearrange? Interchange components? Other pattern? Other layout? Other sequence? Transpose cause and effect? Change pace? Change schedule?

Reverse? Transpose positive and negative? How about opposites? Turn it backward? Turn it upside down? Reverse roles? Change shoes? Turn tables? Turn other cheek?

Combine? How about a blend, an alloy, an assortment, an ensemble? Combine units? Combine purposes? Combine appeals? Combine ideas?

Strategies for Stimulating Solutions

Osborn's checklist is rather versatile in its potential applications. While its main function is to suggest improvements for objects or processes, the seventy-three questions may evoke ideas for other kinds of problems: for instance, in personal, social, educational, scientific, or artistic situations. Osborn himself was in advertising and undoubtedly found many imaginative sales strategies with his list. "Who else instead?" "Other place?" "Othertime?" "Exaggerate?" "Understate?" "Transpose cause and effect?" and other mind-stretchers can arouse a number of creative and probably successful advertising ideas.

Attacking the problem of stopping burglaries from a clothing merchant's storeroom, Parnes (1967) demonstrated how Osborn's questions may be used to modify ideas already suggested:

Listed idea: Put police dogs in storeroom.

New ideas suggested by checklist: *Magnify*—use tiger, put roar of animals on record player; *Minify*—install a parrot trained to say, "Beat it"; *Rearrange*—have roving guard with police dog.

For his "Creative Engineering" course at MIT, John Arnold (Mason, 1960, p. 106) developed a short list of self-questions aimed at improving critical engineering features of commercial products:

Can we increase the *function?* Can we make the product do more things?

Can we get a *higher performance level?* Make the product longer lived? More reliable? More accurate? Safer? More convenient to use? Easier to repair and maintain?

Can we lower the *cost?* Eliminate excess parts? Substitute cheaper materials? Design to reduce hand labor or for complete automation?

Can we increase the *salability?* Improve the appearance of the produce? Improve the package? Improve its point of sale?

In a volume modestly titled *How to Make More Money*, Marvin Small displayed a very extensive checklist of idea-stimulators with examples of how each item could lead to a patentable invention. In part, Small suggested:

Can the dimensions be changed? Larger (economy-size packages, photo enlargements, puffed cereals)? Smaller (hearing aids, tabloid newspapers, pocket flashlight, microfilm)? Longer (king-size cigarettes, typewriter carriage for bookkeeping)? Stratify (plywood, storage pallets, layer cake)? Converge (mechanical artificial hands, ice tongs)? Border (mats for pictures, movable office partitions, room separators)?

Can the quantity be changed? More (extra-pants suits, three stockings)? Less (one-ounce boxes of cereals, ginger-ale splits)? Fractionate (separate packings of crackers inside single box, 16-mm. movie film usable as two 8-mm. films, faucet spray)? Combine with something else (amphibious autos, roadable airplanes)?

Can the order be changed? Beginning (self-starter, red tab to open Band-aids)? Assembly or disassembly (prefabricated articles, knockdown boat kits)?

Can the time element be changed? Faster (quick-drying ink, dictating machine, intercom system)? Slower (high-tenacity yarns for longer-life tires, $33\frac{1}{3}$ rpm long-playing records)? Longer (Jiffy insulated bags for ice cream, wood preservative)? Shorter (pressure cooker, one-minute X-ray machine)? Chronologized (defrosting devices, radio clocks)? Renewed (self-charging battery, self-winding watches)?

Can the cause or effect be changed? Energized (magneto, power steering)? Altered (antifreeze chemicals, meat tenderizer)? Destroyed (tree spraying, breath and perspiration deodorants)? Counteracted (circuit-breaker, air conditioning, filters)?

Can there be a change in character? Stronger (dirt-resistant paint)? Weaker (Pepsi-Cola made less sweet, children's aspirin)? Interchanged (interchangeable parts, all-size socks)? Resilient (foam-rubber upholstery, cork floors)? Uniformity (standards in

Strategies for Stimulating Solutions 111

foods, drugs, fuels, liquor)? More expensive (cigarettes in cardboard or metal boxes, deluxe editions of books)?

Can the form be changed? Animated (moving staircases, package conveyors)? Speeded (meat-slicing machine)? Slowed (shock absorbers)? Deviated (traffic islands)? Repelled (electrically charged fencing)? Admitted (turnstiles)? Lifted (fork-lift truck)? Lowered (ship locks)? Rotated (Waring blender, boring machine)? Oscillated (electric fan)? Agitated (electric scalp stimulator)?

Can the state or condition be changed? Harden (Bouillon cubes)? Soften (Krilium soil conditioner, water softeners)? Preformed (prefabricated housing, prepared Tom Collins mixer)? Disposable (disposable diapers, Kleenex tissues) Parted (caterpillar tractors, split-level highways)? Vaporized (nasal medication vaporizers)? Pulverized (Powdered eggs, lawnmower attachment to powder leaves, garbage disposal)? Lubricated (self-lubricating equipment)? Drier (dehumidifier for cellars, tobacco curing)? Effervesced (Alka-Seltzer)? Coagulated (Jello, Junket desserts)? Elasticized (latex girdles, bubble gum, belts)? Lighter (aluminum luggage, electric blankets)?

Can the use be adapted to a new market? Men (colognes, lotions)? Children (junior-size tools, cowboy clothes)? Foreign (Reader's Digest foreign editions)? [3]

Mr. Small's checklist, which is only partly reproduced here, may not make you instantly rich, as the title of his book seems to imply. At the same time, however, such a list could reasonably stimulate a quantity of fanciful and potentially valuable idea-combinations in the invention-oriented mind of the design engineer or executive.

One of our own creative problem-solving programs, *Thinking Creatively: A Guide to Training Imagination* (Davis and Houtman, 1968; see Chapter 10), teaches sixth-grade to eighth-grade students some effective strategies for designing, inventing, and improving physical products. The

TABLE 8-1 · Idea Checklist (Davis and Houtman, 1968)

CHANGE COLOR?	NEW SIZE?	CHANGE SHAPE?	NEW MATERIAL?	ADD OR SUBTRACT SOMETHING?	REARRANGE THINGS?	NEW DESIGN?
Blue	Longer	Round	Plastic	Make Stronger	Switch Parts	From Other Countries?
Green	Shorter	Square	Glass	Make Faster	Change Pattern	Oriental Design
Yellow	Wider	Triangle	Fiberglass	Exaggerate Something	Combine Parts	Swedish Design
Orange	Fatter	Oval	Formica	Duplicate Something	Other Order of Operation	Mexican Design
Red	Thinner	Rectangle	Paper	Remove Something	Split Up	French Design
Purple	Thicker	5-Sided	Wood	Divide	Turn Backward	Eskimo Design
White	Higher	6-Sided	Aluminum	Make Lighter	Upside Down	Russian Design
Black	Lower	8-Sided	Nylon	Abbreviate	Inside Out	American Design
Olive Green	Larger	10-Sided	Cloth	Add New Do-Dad	Combine Purposes	Indian Design
Grey	Smaller	Lopsided	Gunny Sack (Burlap)	Add New Smell	Other	Egyptian Design
Brown	Jumbo	Sharp Corners	Cardboard	New Sound	Switcheroo?	Spanish Design
Tan	Miniature	Round Corners	Steel	New Lights		
Silver	Other Size?	Egg-Shaped	Leather	New Flavor		From Other Time?
Gold		Doughnut-Shaped	Copper	New Beep Beep		Old West
Copper		"U" Shaped	Rubber	New Jingle		Roaring Twenties
Brass		Other Shapes?	Other Material?	Subtract The Thing That Doesn't Do Anything		Past Century
Plaid			Combination of These Materials?			Next Century
Striped						Middle Ages
Polka-Dotted						Cave Man
Flowers						Pioneer
Speckles						
Paisley						From Other Styles?
Pop Art						Hippie
Other Colors?						Beatnik
Color Combination?						Other Wierdos
Other Patterns?						Ivy League
						Secret Agent
						Elves and Fairies
						Clown
						Football Uniform

Strategies for Stimulating Solutions

workbook therefore instructs students in using checklists as idea sources. The major idea checklist in *Thinking Creatively,* shown in Table 8–1, is a modification of Osborn's (1963) "seventy-three idea-spurring questions." Since we wanted to make very explicit how familiar objects (such as shoes or a popup toaster) may be easily restyled and improved, the checklist provides lengthy sublists of alternative colors and color patterns; possible shapes; different ways to rearrange purposes, parts, and functions; and so on. The section entitled *New Design* (unique to this checklist) presents novel suggestions that should stimulate interesting thoughts for designing clothes, furniture, appliances, and other things —thoughts that probably would never have occurred to the student without the combination-fostering action of the checklist.

We further derived a very brief idea checklist by using just the section headings from the list in Table 8–1. The result is a seven-item list that suggests general categories of solutions for problems of the product-improvement variety: (1) Change Color? (2) New Size? (3) Change Shape? (4) New Material? (5) Add or Subtract Something? (6) Rearrange Things? (7) New Design or Style? Two laboratory experiments (Davis and Roweton, 1968; Warren and Davis, 1969; see also Davis, Roweton, Train, Warren, and Houtman, 1969) confirmed that the seven-item list could stimulate the thinking of college-age Ss.

In the Davis and Roweton (1968) study, Ss were allowed ten minutes to think of as many ways as they could to change or improve a thumbtack (simple object), and another ten minutes to list changes or improvements for a kitchen sink (complex object). Subjects supplied with the short checklist generated two and one-half times the number of ideas listed by Ss in the control (no checklist) group. And what about quality? The average rated "creativity" of ideas was significantly higher in the checklist group than in the control group.[4] Furthermore, the proportion of creative ideas (number "creative" ideas/total) was 25 percent for the checklist Ss, considerably greater than the 13 percent produced by the Control Ss. Object complexity had no effect upon the performance of either group.

A subsequent study (Warren and Davis, 1969) replicated very nicely the effectiveness of the seven-item checklist. This experiment allowed unlimited time to "think of ways to change or improve a doorknob." Also, in addition to a short checklist and a (no-checklist) control group, a third group of Ss were provided with Osborn's "seventy-three idea-spurring questions." The results showed that while Ss working with Osborn's checklist fared no better than Control Ss, short-checklist Ss listed double the number of ideas listed by either of the other two groups. Again, the ideas stimulated by the short checklist were rated significantly higher in quality, although the proportions of "creative" ideas were about equal for the three groups.

Since the Warren and Davis Ss worked as long as they liked, we could obtain two additional measures: S-determined working time, which could be taken as a measure of motivation or interest, and the derived speed measure of ideas per minute. Subjects in the short-checklist group worked longer and produced ideas at a faster rate than Ss in either of the other two groups. Considering both the Davis and Roweton (1968) and the Warren and Davis (1969) experiments, it seems evident that the brief, seven-item checklist—which merely provided general categories of problem solutions—both challenged and stimulated the idea-producing capabilities of the Ss.

The failure of Osborn's well-known list to increase creative productivity in the Warren and Davis study, along with some unsuccessful attempts by Train (1967; Davis, Roweton, Train, Warren and Houtman, 1969) to stimulate product improvements with lengthy checklists, brought home another point: the availability of an idea checklist does not guarantee

Strategies for Stimulating Solutions

its profitable use. The checklist technique can impose an unfamiliar—and with bright, imaginative college Ss, an unnecessary—method of coping with simple tasks. As with all other "forced combinations" procedures, the checklist strategy should be used to extend, not replace, active imaginations.

CHECKLISTS FOR IDEA EVALUATION

In the evaluation of ideas a structured sequence of questions may help distinguish whether or not an idea is a good one. A good idea or problem solution is one that not only works but is worth the effort required to implement it and is attractive to those passing judgment. Writing mainly for a business-oriented audience, Mason (1960, p. 120) proposed these questions to ask about each idea:

Is the idea simple? Does it seem obvious?—or is it too clever? —too ingenious?—too complicated?

Is it compatible with human nature? Could your mother, or the man next door, or your cousin, or the service-station attendant, all accept it?

Is it direct and unsophisticated?

Can you write out a simple, clear, and concise statement of it? Can you do this in two or three short sentences so that it makes sense?

Can it be understood and worked on by people of the average intelligence level found in the field?

Does your idea "explode" in people's minds? Does someone else react to it with "Now why didn't I think of that?" Can people accept it without lengthy explanation? If it does not explode, are you sure you have really simplified it?

Is it timely? Would it have been better six months or a year ago? (If so, is there any point in pursuing it now?) Will it be better six months from now? (If so, can you afford to wait?)

The military also has used criteria listing for idea evaluation. Both of the following lists of questions (from Mason, 1960, pp. 120–122) would lend themselves to problem solving in business, education, or other institutions. From the U.S. Air Force:

Is it suitable? Will this solution do the job? Will it remedy the problem situation completely or only partially? Is it a permanent or stop-gap solution?

Is it feasible? Will it work in actual practice? Can we afford this approach? How much will it cost?

Is it acceptable? Will the company president (or the board, or the union, or the customers) go along with the changes required by this plan? Are we trying to drive a tack with a sledge hammer?

And from the U.S. Navy:

Will it increase production—improve quality?

Is it a more efficient utilization of manpower?

Does it improve methods of operation, maintenance, or construction?

Is it an improvement over the present tools and machinery?

Does it improve safety?

Does it prevent waste, or conserve materials?

Does it eliminate unnecessary work?

Does it reduce costs?

Does it improve present office methods?

Will it improve working conditions?

According to the Navy, if the answer to any *one* of these questions is "Yes," the idea is judged constructive (Mason, 1960, p. 122).

Strategies for Stimulating Solutions

POLYA'S CHECKLIST FOR MATHEMATICAL PROBLEM SOLVING

Apart from checklists that stimulate imagination and lists of criteria for evaluation, Polya (1957) presented a mathematical problem-solving checklist. Instead of eliciting a large number of solutions, as in the case of idea checklists, Polya's list teaches different forms of questioning geared to defining and planfully (with imagination) approaching difficult and unfamiliar mathematical tasks. With little modification, most of his principles and questions are easily adaptable to problem solving in general:

First, you have to understand the problem.

Understanding the problem: What is the unknown? What are the data? What is the condition? Is it possible to satisfy the condition? Is the condition sufficient to determine the unknown? Or is it insufficient? Or redundant? Or contradictory?

Draw a figure. Introduce suitable notation.
Separate the various parts of the condition. Can you write them down?

Second, you should obtain eventually a plan of the solution. You may be obliged to consider auxiliary problems if an immediate connection cannot be found.

Devising a plan: Have you seen it before? Or have you seen the same problem in a slightly different form?

Do you know a related problem? Do you know a theorem that could be useful?

Look at the unknown! And try to think of a familiar problem having the same or a similar unknown.

Here is a problem related to yours and solved before. Could you use it? Could you use its result? Could you use its method?

Should you introduce some auxiliary element in order to make its use possible?

Could you restate the problem? Could you restate it still differently? Go back to definitions.

If you cannot solve the proposed problem try to solve first some related problem. Could you imagine a more accessible related problem? A more general problem? A more special problem? An analogous problem? Could you solve a part of the problem? Keep only a part of the condition, drop the other part; how far is the unknown then determined, how can it vary? Could you derive something useful from the data? Could you think of other data appropriate to determine the unknown? Could you change the unknown or the data, or both if necessary, so that the new unknown and the new data are nearer to each other? Did you use all the data? Did you use the whole condition? Have you taken into account all essential notions involved in the problem?

Third, carry out your plan.

Carrying out the plan: Carrying out your plan of the solution. Check each step. Can you see clearly that the step is correct? Can you prove that it is correct?

Fourth, examine the solution obtained.

Looking back: Can you check the result? Can you check the argument? Can you derive the result differently? Can you see it at a glance? Can you use the result, or the method, for some other problem? *

The checklist technique for stimulating new idea-combinations, for suggesting new problem approaches, and for evaluating ideas against carefully enumerated criteria, is largely self-explanatory and requires little further comment. The essence of checklisting is simply that when stuck for new concepts and viewpoints, an appropriate checklist may very well

* From G. Polya, *How to Solve It: A New Aspect of Mathematical Method* (copyright 1945 by Princeton University Press, © 1957 by G. Polya; Princeton Paperback, 1971), pp. 6, 7, and 8. Reprinted by permission of Princeton University Press.

trigger some valuable idea-combinations that would otherwise never have occurred to the thinker.

Summary: Comment on "Forced Creativity"

The attribute-listing strategy of Crawford and the similar morphological synthesis method of Zwicky and Allen are no more, but no less, than they appear. They are simple techniques for producing new combinations of ideas—so simple, in fact, they can be learned by elementary school students. Do these forced idea-generation techniques represent "true" creative thinking and problem solving? The authors of these procedures —pointing to instances of attribute listing and "morphologizing" in the real world of creative behavior—say Yes. At the very least, we must agree that (1) such procedures can lead to new, potentially valuable ideas and problem solutions, and (2) even the most intuitive of creative persons may find himself modifying attributes, forcing combinations, and hunting for ideas on checklists.

NOTES

1. Crawford also wrote *Direct Creativity* (1964), *How to Get Ideas* (1948), and *Think for Yourself* (1937).
2. Laboratory research testing the effectiveness of the morphological synthesis technique is almost nonexistent. One study by Warren and Davis (1969) showed that the technique can indeed stimulate the imaginations of college students.
3. The complete checklist is available in Simberg (1964, pp. 127–136).
4. Ideas rated above the midpoint of a seven-point creativity scale by each of two judges were considered "creative."

◄ CHAPTER 9 ►

METAPHORICAL THINKING AND PROBLEM SOLVING: SYNECTICS AND BIONICS

METAPHORICAL THINKING lies at the heart of creative writing. Indeed, one mark of every outstanding author or poet is his genius for colorful, metaphorical word play. It also is true that metaphorical and analogical thinking may be used deliberately in problem solving, as when we transfer a solution from an old problem to some new, analogous problem context. For example, if we tried to design a basement ceramics kiln for firing clay pottery, our main difficulty is "How can we heat it?" We might find ideas by analogically considering how other heating problems are solved—perhaps transferring an electrical, butane, or acetylene solution to our own kiln-heating problem. Aside from the relatively sober notion of transferring solutions from related problems, two sets of techniques—the synectics methods and bionics—deliberately make use of often farfetched analogical thinking to produce highly creative problem solutions.

Synectics

Synectics, taken from the Greek *synecticos,* meaning the joining together of apparently unrelated elements, is many things—all of them creative. At present, synectics is a corporation, Synectics, Inc., whose three main functions are (1) solving thorny design-engineering problems for commercial clients, (2) training professional problem-solving groups, and (3) developing inner-city educational materials based upon metaphorical thinking.[1] Synectics also is a set of specific strategies for producing imaginative ideas and problem solutions and a problem-solving group leadership technique which differs drastically from the brainstorming leadership function. Above all, synectics is a state of mind disciplined for deliberate flexibility and imagination.

THE ORIGINATOR: WILLIAM J. J. GORDON

In the above "synectics is" paragraph, I could have added that synectics is William J. J. Gordon, originator of the concept. A brief sketch of Gordon's something-less-than-dull background might run as follows:[2] His college career included stints at the University of Pennsylvania, University of California, Boston University, and Harvard, where he received his A.B. degree. Setting aside his training in history, physics, psychology, and philosophy, Gordon sought to apply his biochemical coursework to breeding pigs for bacon, controlling hormones and vitamins. The outcome, according to Gordon, was ". . . a lot of bone and not much bacon, but they were the fastest pigs in the East" (Alexander, 1965, p. 168).

After the pigs, Gordon broadened his experience by working as master of a sailing schooner, schoolteacher, horse-handler, ski instructor, inventor, author, college lecturer, and in World War II, an ambulance driver and salvage diver off North Africa. Finally, in a position leading to his interest in the hows and whys of creative group problem solving, Gordon served as a research associate with a Harvard group whose charge was to develop an acoustic torpedo.

The work with the Harvard team, and with the still-existing Invention Design Group that he established in 1952 at Arthur D. Little, Inc., convinced Gordon that hidden processes of creative problem solving may be both understood and accurately described. Even more important, these creative processes or mechanisms may be taught to others to aid them in problem solving—with more than a touch of imagination.

THE SYNECTICS MECHANISMS: STRATEGIES FOR
CREATIVE PROBLEM SOLVING

We have mentioned that the synectics mechanisms or strategies are based in large part upon metaphorical thinking. Three of Gordon's pet analogical devices for stimulating new problem approaches are the *direct analogy, personal analogy,* and *fantasy analogy* methods (Gordon, 1961).

When William Gordon or Synectics, Inc., president George Prince announce to their problem-solving group members, "O.K., let's find a direct analogy," the leader and followers begin looking for remote problem parallels, especially from natural biological systems. Such metaphorical thinking—asking how animals, birds, insects, flowers, or trees have solved similar problems—provides the combinatory play which, from experience, they know will lead the group to some rather imaginative new viewpoints.

For example, in one session the group tackled the problem of designing a new roof that would turn white in summer, to reflect heat, and black in winter, to absorb heat (Gordon, 1961, pp. 54–56). The opening question asked for a direct analogy: "What in nature changes color?" After discarding a weasel, the group landed on a flounder, a sea-bottom fish that adjusts its light-dark coloration to match its surroundings. The color-changing devices are the chromatophores, tiny sacs containing black pigment which are connected to the spinal cord. The chromatophores contract to push the black pigment toward the surface, functionally darkening the fish's exterior. Conversely, relaxing the chromatophores retracts the pigment and lightens the tone.

Now, as one synectics member put it, "Let's flip the flounder analogy over to the roof problem" (p. 55). The final viewpoint was an all-black roofing material impregnated with "little white plastic balls" which would expand when the roof became hot, changing the surface to a heat-reflecting white. A drop in temperature would contract the white balls, returning the roof to its heat-absorbing dark color.

The personal analogy method—imagining oneself to be a problem-object—also seems to guarantee an original point of view. In the transcript of one imaginative session (Prince, 1968), the members tried to devise a way to bring up representative samples of oil-saturated rock from beneath a reservoir. The difficulty was that during ascension, pressure changes and sloshing through muddy water probably (they never knew for sure!) distorted the samples before they reached the surface. Since the problem was imaginatively redefined as determining the crowdedness of the oil in the rock, the synectics thinkers suggested for exploration the direct analogy (from biology) of a crowded virus culture. The intent, as always, was to examine the problem from unusual metaphorically-related directions.

The leader asked for personal analogies: "Let's take this virus culture. Now take a couple of minutes . . . don't say anything while you get into your new skin. . . . You are one virus in this culture. . . . How do you feel?"

With little immediate concern for propriety or relevance, each of several group members freely described his new form, functions, emotions and attitudes.

ONE MEMBER: I am very small . . . curled like a corkscrew.
SECOND MEMBER: It's nice and warm in this culture, but I feel itchy.

THIRD MEMBER: I feel a sense of real urgency . . . panic because I keep turning into two of me and then four. . . . I can feel the food getting hard to get . . . got to die.
FIRST MEMBER: I hate the world. I want to get . . . out of this culture where I can kill other things alive. It's a black world. I want to murder.
FOURTH MEMBER: I feel I am a successful virus. With the way these guys feel I can sit back and relax, enjoy life and play a guitar.
THIRD MEMBER: I resent his playing his guitar while I'm panicky!

Obviously, the members of the group became quite involved in the role-playing. But did they solve the problem? In this particular session, the virus personal analogy itself did not lead directly to a solution or viewpoint. But by a rather circuitous route the group eventually considered a raging cat and proposed to calm the cat by stroking or cooling or freezing it. Returning to the problem, the now-obvious insight was to freeze the rock sample (perhaps by pumping down liquid nitrogen) so that it would not change form or content on the trip to the reservoir surface. The group thus found a realistic, creative solution via a remarkable sequence of metaphorical thinking, an example of uncensored tapping of our immense storehouse of unconscious information (Prince, 1968).

In addition to the direct analogy and personal analogy mechanisms, Gordon (1961) proposed the fantasy analogy strategy, a form of Freudian wish-fulfillment. The user of fantasy analogy searches for ideal-but-farfetched solutions. Asking how a problem could solve itself, for example, is an ideal approach which may lead to effective problem solutions. To illustrate, years ago someone probably asked how to make a refrigerator defrost itself, how to make tires repair their own leaks, and how to make an oven clean itself. Such ideal solu-

Metaphorical Thinking and Problem Solving

tions probably led to the invention of self-defrosting refrigerators, self-sealing tires, and self-cleaning ovens.

As Gordon (1961) uses fantasy analogy, the method is even more fanciful than speculating on how your problem can solve itself. In one session, the task was to invent an airtight zipper for space suits. In response to the opener, "How do we in our wildest fantasies desire the closure to operate?" (Gordon, 1961, p. 49), group members imagined an insect, working on command, running up and down the closure manipulating little latches. Eventually, the insect (fantasy analogy) ideas, interspersed with a few direct analogies and personal analogies, led to a simple and novel vaporproof closure device: each side of the space-suit opening contained a long spring embedded in rubber. When the two sides were pressed together, the convolutions of the two springs overlapped; a wire inserted through the overlapping portion of the springs kept the two sides pulled tightly together.

The mechanisms of direct analogy, personal analogy, and fantasy analogy make conscious use of some conceptually elusive creative abilities and abstractions: the use of irrelevance, detachment, free metaphorical word play, empathy, and intuition. While we hardly can teach or even adequately describe such concepts, the synectics mechanisms that elicit empathy, detachment, metaphorical behavior, etc., are in fact sufficiently concrete to be both described and taught. Further, it is very important to note that the synectics approaches to problem solving almost entirely eliminate the common problem-solving bugaboos of habit, fixation, and conformity.

Before describing the relatively sophisticated functions of the synectics group leader, we should mention that the flexibility-creating synectics mechanisms may be applied to many types of problems, in individual or group thinking sessions, and without a highly trained synectics leader. Based upon our own efforts to teach problem solving by analogy (Davis,

1969; Davis, Houtman, Warren, and Roweton, 1969), there is good evidence that sixth-grade to eighth-grade students can learn to use slightly simplified forms of the synectics methods. That is, they can learn to find ideas by asking how animals, bugs, and maybe even plants have solved similar problems, or by searching for perfect (or ideal) types of problem solutions (Davis and Houtman, 1968, pp. 15, 85).

THE SYNECTICS LEADERSHIP FUNCTION

In professional synectics group problem solving, the recommended leadership strategy is fairly complex. Unlike Osborn's (1963) brainstorming sessions, where all ideas are freely tossed out for later (deferred) evaluation, the synectics leader skillfully directs the entire show. He makes numerous important decisions as to which definition of the problem to adopt, which mechanisms (direct, personal, or fantasy analogy, or some other idea-stimulating device) to use, and which line of thought to follow. We will see that the sequence of problem-solving stages through which the synectics leader guides his group bears a remote, but only remote, resemblance to the traditional *preparation, incubation, illumination,* and *verification* stages of Wallas (1926; see Chapter 2).

The steps: Prince (1968) labeled the first step in synectics group problem solving as *problem as given* (PAG). As the name suggests, the PAG is a statement of the problem as initially presented by an outside source. In the reservoir problem, the PAG was simply, "How do we determine oil saturation in reservoir rock?"

Second, the group hears a detailed *analysis* of the problem, preferably by an expert.

Purge, the third step, is the immediate airing of obvious solutions—so the expert, who probably has thought of them already, can explain why they will not work. Everyone now feels better, and they understand the problem more thoroughly. On rare occasions, a *purge* idea may solve the problem.

Fourth, each participant rewrites the *problem as given* (PAG) as he would like to view it, proposing a *problem as understood* (PAU). The PAU stage already may reflect an imaginative hunt for new perspectives. In the reservoir problem two PAUs were "How to make the reservoir rock tell the truth" and "How to have oil tell me how crowded it is in the reservoir rock" (Prince, 1968, p. 10).

Fifth, the leader selects a PAU, then starts the metaphorical activity by asking an *evocative question* (EQ). He may ask for a direct analogy, a personal analogy, or a fantasy analogy. According to Gordon (1961), it is best to begin with a fantasy analogy (again, a wish-fulfilling "ideal" problem solution), which will induce direct and personal analogies. That is, stating an ideal, even though fantastic, solution will stimulate the search for parallel solution mechanisms from nature (direct analogies) and likely some imaginative role-playing of the personal analogy variety.

At this stage of the evocative question, the session loosens up. The leader now is faced with the task of choosing which analogy (virtually a line of free association) to follow, when to ask another evocative question (for a different type of analogy), and when to move to the semifinal step, *force fit*.

In the sixth step, force fit, the group deliberately tries to transfer the substance of the analogy of the moment to the real problem. In the reservoir-rock problem, the leader announced at an appropriate point, "Let's move into *force fit*. How can we take this idea of cat and use it to help us have the oil tell us how crowded in the reservoir rock it is?" (Prince, 1968).

Seventh, from the force fit, the members move toward the final stage, the *viewpoint*. The viewpoint may be a solution or a novel approach to the problem. As we mentioned above, in the reservoir problem one written-up viewpoint was freezing the reservoir rock before bringing it to the surface.

The viewpoint, however, does not end the session. Successive "excursions" with a new problem as understood, new evocative questions, or new analogies result in further viewpoints for the same problem.[3]

In regard to our *attitudes, abilities,* and *techniques* model of creative problem solving, synectics nicely demonstrates all three components. As with brainstormers, favorable attitudes toward uncensored, flexible thinking are the *modus operandi* of experienced synectics groups. The synectic mechanisms, direct analogy, personal analogy, and fantasy analogy, are precisely conscious and deliberate techniques for producing highly imaginative idea-combinations. Finally, synectics problem solving makes use of, and very likely strengthens, such abilities as detachment, empathy, flexibility, originality, and the complex abilities to think metaphorically and to role play.

Bionics

Bionics has been defined as ". . . the use of biological prototypes for the design of man-made systems" (Papanek, 1969), or similarly, ". . . the study of the structure, function, and mechanisms of plants and animals to gain design information for analogous man-made systems" (Advanced Technology Staff, Martin Company, 1963a.)[4] It is, in short, a problem-solving strategy, resembling the synectics direct analogy method, whose infinite storehouse of ideas is nature itself.

The bionicists comprising the Advanced Technology Staff of the Martin Company (1961, 1963a, 1963b) are specialists in both the biological and engineering sciences. Their biology-based design engineering strategy is to examine closely

Metaphorical Thinking and Problem Solving

the motor, circulatory, neural, and especially the sensory capabilities of organisms from a variety of suburbs of the animal kingdom—mammals, birds, reptiles, amphibians, fish, and insects. The ambition of each bionicist—not necessarily a modest one—is to achieve an occasional breakthrough in such engineering goals as increasing reliability, sensitivity, strength, maneuverability, or speed, while reducing size, weight, or power requirements. On all counts, the bionicist can easily point to biological systems which overwhelmingly outstrip any manmade analog.

Sensory organs, as a class of transducers, detect mechanical, thermal, electrical, or chemical energy, transforming it into nerve impulses. A number of incredible animal-sensing mechanisms have suggested to the prepared bionicist parallel electronic devices for solving some tricky engineering problems. For example, studying the frog's eye—which responds to objects moving against a still background (potential bugs)—led to (1) a telephone filter and (2) a radar scope, both of which suppress background noise while amplifying the active signal. The eye of a beetle, which integrates shadow movements into ground speed and directional information, suggested a similar aircraft device: with one photo cell in the aircraft nose and another in the tail, detection of the same light pattern reflected from earth provides precise ground-speed information. Also, an adaptation of the device may be used to advance film accurately in airborne cameras.

It is common knowledge that bats possess a form of sonar. The efficiency of the sonar is astounding: it can pick up echoes from a mosquito against a background two thousand times stronger than the signal. One variety of bat can locate fish with his specialized sonar gear. Furthermore, the bat's auditory equipment weighs a fraction of a gram. There is ample reason why designers of improved sonar—for example, the U.S. Navy—study bats as well as electronics.

While the bat is busily devouring most flying insects, he

may experience some difficulty nabbing the average moth, whose simple two-celled ear will uniquely separate bat sounds from background noise.

Similarly amazing, the infrared sensors of the rattlesnake respond to a heat change of .001° C. in .1 second—which might be the difference between his sun-heated rock and a rabbit. I confess I cannot judge whether or not mankind is better off, but the rattlesnake infrared sensor led to the appropriately named *Sidewinder* missile, a device which indifferently trails the hot exhaust of a target aircraft, exploding in its tailpipe.

Some varieties of fish are sensitive to electric potentials no stronger than the earth's magnetic field, enabling them to detect alterations in the electric field caused by objects (such as a comb recently run through the bionicist's hair) or other fish nearby. Whales and porpoises have very high propulsion efficiency, allowing them to slide through water with noticeably less friction than the average boat or submarine. The human body regulates itself within 1/100 of a degree Fahrenheit, despite outside temperature, maintaining the familiar 98.6°. Such natural engineering feats justifiably intrigue the bionics-oriented engineer.

One familiar analogy is between the computer and the human brain: both possess complex forms of information input, storage, processing, and retrieval. This computer-brain correspondence actually is studied from both ends: analyses of brain structure and process may suggest innovation in computer design, and studying computer "behavior" may shed light upon human neurocognitive processes (Papanek, 1969). Following the former strategy, Raytheon's *Cybertron,* constructed to duplicate human learning processes (for example, trial and error, correlating new facts with experience) can reliably distinguish sonar signals bounced off submarines from those bounced off porpoises after a few hours of "training." Humans require months (Advanced Technology Staff,

Martin Company, 1961). Following the latter strategy, scientists at the Cornell Aeronautical Laboratory study self-organizing pattern-recognition programs, which find key invariant properties while rejecting all others, to help understand the structure and function of the human brain. Also, in Chapter 5 we saw that a computer program which solves problems will specify in detail a sequence of steps that a human may analogously follow when working on the same problem.

Conclusion

Undoubtedly, nature's highly efficient "machines" can provide new viewpoints and solutions to a large assortment of metaphorically related human problems. While the examples mentioned in the bionics section would interest mainly the professional systems-design engineer, the general strategy of looking to biology for innovative problem solutions could prove fruitful in a variety of problem-solving contexts. For example, auto safety experts might look at anticollision devices displayed by myriads of everyday organisms. It is noteworthy that the direct analogy method—which, like bionics, directs the thinker to seek new ideas and viewpoints in nature—is the core idea-finding mechanism in synectics problem solving. There is no limit to the number of ideas and devices waiting to be adapted.

NOTES

1. Three workbooks—*High School Social Studies Text;* an invention-focused vocational program, *The Art of the Possible;* and the language-arts materials *Making It Strange*—have re-

cently been produced by Synectics Educational Systems. The theory and classroom application of parts of these workbooks appear in *The Metaphorical Way of Learning and Knowing* (Gordon, 1969), also available from Synectics Educational Systems.
2. Details are available in Alexander (1965), a delightful *Fortune* magazine article headed by an Al Capp portrayal of a typical synectics group-think session. The article and cartoon are reprinted in Davis and Scott (1971).
3. This sequence of steps is a slightly simplified version of the synectics stages present by Prince (1968).
4. Reprinted in Davis and Scott (1971).

◄ CHAPTER 10 ►

IMAGINATIVE PROBLEM SOLVING IN THE CLASSROOM: THE WISCONSIN PROJECT

THE WISCONSIN PROJECT for training creative problem solving in the schools evolves about our two concepts of creative attitudes and creative thinking techniques. Both are critical for creative behavior, and both are sufficiently concrete to be taught to others.

Creative Attitudes: Imaginative Problem Solving As a Voluntary Act

One of our basic assumptions is that using one's imagination in creative problem solving is in part a voluntary act. Anyone capable of reading these words is more than capable of stretching his mind in a completely deliberate manner. He can voluntarily think of wild, never-experienced things such as the details of some supercity on an imaginary planet. The longer he thinks, that is, the more effort he invests, the greater the quantity and the novelty of the ideas—the inhabitants (people or what-have-you), their clothing (if any), their vehicles (or other transportation devices), their shelters, streets, and community buildings, even the details of their superkitchens. Returning to earth, the reader can also deliberately think of new, perhaps workable solutions to today's real problems of war, poverty, traffic safety, drug abuse, air and water pollution, and the ever-mushrooming world population.

In our own laboratory research we found that college students easily and voluntarily could adjust their creativity test responses—responses that, incidentally, are supposed to reflect a stable cognitive ability. When asked to "be original" in thinking of uses for a tire, screwdriver, or coat hanger, the students gave original answers. When the same students were instructed to "be practical," their ideas became practical; and when instructed to "be wild," their responses became more unique than ever (Manske and Davis, 1968).[1] Wallach and Kogan (1965) also concluded that highly intelligent, but not highly creative, students were disinclined, rather than unable, to use their imaginations.

Throughout this book, we have used the term *attitudes* to describe the individual's predisposition to regularly and voluntarily use his creative imagination. While the term *attitude* normally implies a positive or negative liking of some sort —as when we speak of favorable or unfavorable attitudes toward some object, person, or idea—we have extended the definition of attitudes to include sets, predispositions, and awarenesses. Our three sets of training materials described below seek to teach a few pivotal attitudes aimed at increasing the likelihood of creative problem-solving behavior by students.

The most important of these attitudes is simply a strong, positive orientation toward new and innovative ideas. This favorable disposition toward farfetched ideas includes a set to produce imaginative solutions and a receptiveness to the innovative ideas of other thinkers. In professional creative problem solving, we have seen that the single most important principle of brainstorming is deferred judgment, the encouragement and acceptance (for possible taming down) of even the wildest suggestions. The *synectics* metaphorical problem-solving methods encourage wild associative play even more than in brainstorming. In any field or media, innovations that are socially recognized as creative are, by definition, new and imaginative. There is no reason that this simple attitudinal principle—an appreciation for imaginative new ideas—cannot be taught in the schools as it is (or should be) in college and in professional creative problem-solving courses.

Another attitude conducive to creative behavior has been named *constructive discontent* (Osborn, 1963), referring to the notion that any manmade object or process may be changed for the better. "Any" opens a number of possibilities. Guilford's (1967) concept of problem sensitivity, measured by tests which ask for improvements for several appliances or several institutions, overlaps with the concept of constructive discontent. Incidentally, we would prefer to call constructive discontent (or problem sensitivity) an attitude rather than an ability for the semantic reason that attitudes are considered to be more teachable than are genetically determined abilities.

Also included in our broadly conceived definition of creative attitudes is an awareness of the importance of new ideas in all aspects of our fast-changing society. More directly relevant to students, we emphasize that their capability to produce new ideas will tremendously aid them in their own future occupations, whether in business, the arts, teaching, professional sports, or any other occupation (Davis and Houtman, 1968).

Another important yet teachable attitude, geared toward the student's self-concept, is simply the notion that we can learn to be more imaginative problem-solvers—with interest, effort, and perhaps a little guidance.

Creative Thinking Techniques

In addition to the attitude-based voluntary nature of at least some degree of creativity, we further assume that a working knowledge of such thinking strategies as attribute listing, morphological synthesis, idea checklists, and the metaphor-based synectics and bionics can improve one's potential for imaginative problem solving. That is, the conscious use of these techniques may substantially supplement the "unconscious" or intuitive use of unspecified strategies.

Wisconsin Project: Three Workbooks

Our concepts concerning creative attitudes and creative problem-solving techniques have been incorporated in three sets of materials for stimulating imagination in the upper elementary and lower junior high school grades. All three programs seek to encourage creative and problem-solving behaviors (1) by instructions and illustrations dealing with creative attitudes and techniques, (2) by student exercises that allow him to find new ideas while practicing the strategies, and (3) by example, in that the story lines and content of the three materials—especially the *Saturday Subway Ride* —were intended to demonstrate flexibility and imagination.

"THINKING CREATIVELY: A GUIDE TO
TRAINING IMAGINATION"

Our first effort to structure creative problem-solving principles was *Thinking Creatively: A Guide to Training Imagination* (Davis and Houtman, 1968; see also Davis, 1969). Prepared for sixth-grade, seventh-grade, and eighth-grade stu-

The Wisconsin Project

dents, the 150-page workbook takes the form of humorous dialogue among four cartoon characters. Mr. I, an eccentric backyard scientist-inventor (Figure 10–1) teaches the other

Figure 10–1 · Mr. I (Davis and Houtman, 1968).

three characters creative attitudes and various problem-solving techniques. He often engages in activities consistent with a flexible, creative atmosphere, such as fixing a carrot sandwich, pumpkin cake, or some other surprising delight. Dudley Bond, a distant relative of a very famous secret spy (Figure 10–2), is our eager young male character. While a bit awkward at times, Dudley displays a fine sense of humor and enjoys the fascination and challenge of digging up new ideas for solving problems. Maybelle, a curious but slightly naïve young lady (Figure 10–3), is Dudley's friend who needs help in learning to find zesty ideas for her English themes. As Maybelle explains, "I get tired of writing about Hambone, my pet iguana." Last, but hardly least, is Max (Figure 10–4), a professional bear who, being the clown of the program, rarely understands anything very clearly. He often displays his uncreative mean streak and freely criticizes some of the "nutty" ideas—allowing the others (and the authors) frequent opportunity to reemphasize the important creative attitudes.

Figures 10–2 and 10–3 · Dudley and Maybelle (Davis and Houtman, 1968).

Throughout the program, the four friends attack many simple and complex problems. Mr. I explains the creative procedures and attitudes likely to aid in solving a given problem, and Dudley, Maybelle, and sometimes Max use the principles to produce innovative problem solutions. At the

Figure 10–4 · Max (Davis and Houtman, 1968).

end of each of the ten chapters, important principles are reviewed and exercises are presented that allow the students themselves to solve problems similar to those solved by the cartoon characters within the story line.

The program recognizes that strong pressures exist that inhibit a free flow of ideas that might be judged "ridiculous" or "silly." Therefore, to create an atmosphere conducive to relaxed spontaneity, where the wildest ideas may be freely suggested, *Thinking Creatively* deliberately uses humor. The story characters readily propose outlandish problem solutions, and at the same time, engage in slapstick comedy. In addition to contributing to a creative atmosphere, the humor performs an important motivational function, helping to maintain student interest by entertaining them.

The creative problem-solving content of the program focuses upon the creative attitudes described above and such idea-stimulating techniques as attribute listing (renamed the part-changing method), the morphological synthesis procedure (relabeled checkerboard method), checklisting, and many of the metaphor-based synectics methods (one of which, direct analogy, we renamed the find-something-similar method).

In addition, we sought to increase students' understanding of problem solving by describing four problem-solving steps: First, one must clearly understand the problem and define it in general terms. For example, one problem was originally stated as "Think of ways to give your pet hamster a bath. He thoroughly dislikes water, to say nothing of soap!" (Davis and Houtman, 1968, p. 134). Stated more generally, ". . . to open our minds to more kinds of solutions," the real problem is ". . . to separate the dirt from the hamster." The second step is to think of different approaches to solving the problem. In the hamster problem, vacuuming would constitute one approach; soap-and-water tactics would be another. Third, one thinks of different specific ideas for each problem-approach: for example, different ways to vacuum and different ways to apply soap-and-water to the hamster (or the hamster to soap-and-water). The fourth step is to choose the best ideas.

Regarding the effectiveness of *Thinking Creatively,* ideally we would hope for long-term improvement in problem solving and creative skills in the student's personal and educational life, and eventually in his chosen career. Unfortunately, such criteria are not immediately available, and so in our two field evaluations we necessarily resorted to attitude surveys, divergent thinking tests, and informal student and teacher reactions. Briefly, a small pilot evaluation with middle-class seventh-grade students showed that twenty-three students using the program produced 65 percent more ideas on three divergent thinking tasks (ideas rated as significantly more "creative") than thirty-two other students who were enrolled in a creative writing course. The attitude survey indicated that the experimental students were significantly more confident of their creative ability, more appreciative of unusual ideas, and more aware of the importance of creative innovation in society (Davis, Houtman, Warren, and Roweton, 1969). A larger field test with inner-city students (two sixth-grade and two eight-grade classes) showed less, but still reliable, improvement in creative attitudes and idea-producing capabilities (Davis, 1970a, 1970b; Davis, Houtman, Warren, Roweton, Mari, and Belcher, 1972). With the exception of one teacher, whose classroom discipline problems prevented beneficial group problem solving and discussion, all teachers felt the program provided beneficial information and experiences—for students and teachers. It was reinforcing to the authors to learn that one class, at the students' suggestion, used the checkerboard (morphological synthesis) method to generate theme ideas for the entire class; they also used the checklist method to find ideas for decorating the room in an Oriental mode.

"SATURDAY SUBWAY RIDE"

Two considerations regarding *Thinking Creatively* led to the creation of another creative thinking and problem-solving program, *Saturday Subway Ride* (DiPego, 1970). First of all, since *Thinking Creatively* drew heavily from industrial creative problem-solving principles, creative writing was touched very lightly, and such areas as music, artwork, entertainment, and social problems had been left out entirely. Second, there was reason to believe that a highly skilled group of professional creative writers might be better able to teach creative thinking by example, that is, by demonstrating what "being creative" is like.

Like *Thinking Creatively*, *Saturday Subway Ride* is not tied to any one subject area. Rather, both sets of materials teach general principles of flexible thinking and problem solving that are relevant to creative attitudes and idea-finding strategies. However, *Saturday Subway Ride* differs from its predecessor in several respects. For instance, its audience could reasonably be "students" of any age, beginning with fifth or sixth grade, or even lower grades with the aid of a clever and interested teacher. And again, the scope of *Saturday Subway Ride* is considerably broader than *Thinking Creatively*.

The content and flavor of *Saturday Subway Ride* might best be presented in a few illustrative passages:

Let me tell you about last Saturday.

I took a ride on a new super subway that travels a fast circle from Kansas City to Pittsburgh to Dublin to Tokyo to Santa Monica and back.

What's wrong?

You say there's no such subway, and you're about to close the book and stare out the window?

Well, maybe you're wrong. Maybe I zipped around the world on an underground thought, a daydream, a nightdream, or a superfastspecialfivecityidea.

That's what this book is all about.

Ideas.

You say my subway ride is just a wild idea and pretty silly, and you'd rather pitch pennies?

Well, what about flying? People said that men flying around in machines was a wild idea and pretty silly. Then the Wright brothers took off and ZIP!

People once thought that TV was just a wild idea and probably wouldn't work—and bicycles, too, and life insurance and polio vaccine.

A wild idea is something that people find hard to accept because it's new and sounds strange and looks funny and maybe it's light green suede and smells of paprika. Anyway, it's something people haven't seen before, and that makes them afraid.

Some people only feel safe with old, comfortable, tried-out ideas. I guess those people never learned to stretch their minds.

That's what this book is all about, too—learning to stretch your mind, learning to reach out for big, new, different, and even wild ideas.

Why?

So you can solve problems and create new things and improve old things and have more fun. Ideas are good anywhere, anytime, in any climate and even underwater.

Now, you've gotten me way off the subway track. Let's get back to Saturday.

Last Saturday I took a ride on a super subway.

I bought my ticket in Kansas City, Mo., but they wouldn't take money.

"What can you do?" the ticket-seller asked me.

"What do you mean?"

"In order to ride this subway, friend, you have to do something musical. Can you sing?"

I can't sing. I can't dance very well either. And I only had two months of lessons on the B-flat clarinet before I gave it up. I couldn't think of anything musical I could do. But I really wanted that ticket to Pittsburgh.

"Well, can you write me a song?"

"You mean make something up?"

The Wisconsin Project

"Yes," he said.

"Now?"

"Well . . ." He scratched a sideburn and said, "tell you what. You can get on the subway and ride to the first stop—Pittsburgh. When you get to Pittsburgh, you better have a song ready or we'll have to send you right back here." . . .

"How long before we get to Pittsburgh?" was my next question.

"That depends," he said. "We'll get to Pittsburgh as soon as you can lean back in your seat and close your eyes and not think about Pittsburgh. As long as you're thinking about Pittsburgh, we won't get there. Could take days or minutes." . . .

"Well, I can't help thinking about having a *song* ready for Pittsburgh. How do you write a song?"

He looked very thoughtful and squinty and smart and said to me, "I don't know."

I thanked him for being honest. Some people never admit they don't know something. "Excuse me." I left my seat and walked up the aisle.

I met an old woman with blue hair and a radio cane. "That's a good idea," I said, "a transistor radio built right into your cane."

"I'm an inventor," she said. "I'm paid by the job with an advance on royalties based on thousands of items sold. What do you need?"

"A song."

"Oh." She nodded and smiled. "Well, that has to come from your own mind," she said, and turned off her cane.

"But I don't even know where to start."

"Start with rhythm," she said. "A song has to have a beat."

"Where do I find it?"

She only said, "Shhhhhhhhh."

I shut up. I even closed my eyes. But all I heard were the wheels of the subway:

                clackaty clackaty chugity
                clackaty clackaty chugity
                clackaty clackaty chugity

"Clackaty clack . . . hey! That's the rhythm I need!"

She said, "Yes, now comes the melody. Why don't you borrow one, just for now?"

"Like 'My Darlin' Clementine'?"

"Yes, or 'Old Man River.'"

We settled on "You Are My Sunshine."

"Whatever you need for a song is around you and inside of you all the time," she said.

Well, I knew I needed words, so I looked around. Above the subway windows were advertising posters. You've seen them—ads for shaving cream, hair lotion, insurance, concerts, how to get a job, where to go to electronics school, etcetera.

Well, when you're desperate for a song, why not? Here it goes—to the tune of "You Are My Sunshine," and to the clackaty clackaty beat of the subway.

> (CLACKATY CLACKATY CHUGITY)
> Don't drop out of school
> Don't drop out of school
> Remember al-ways
> To look your best
> Elect Polinski
> As County Sheriff
> And the zoo will o-pen in May.

Well, what do you think? Oh, it's not that bad. It might get me to Pittsburgh.

Okay, you try it. But take your time. Maybe you can make your words really mean something, tell a story, or just tell how you feel, or what kind of day it is outside the window. Anyway, take a stab at it. And remember what this book is about—stretching your mind, reaching out a little, bending your brain. Go ahead. What you need to write a song is around you and inside of you all the time. . . .

"Hey! Wow! Yow! Hey!"

It was the conductor, screaming and running down the aisle, his eyes wide, his hat off.

"Hey! Eeeeyah!"

People got their tickets ready in a hurry, figuring the conductor was awfully excited about his job.

The Wisconsin Project

"Gaa! Yow! Wow!"

But he wasn't collecting tickets, just running and yelling. "Wow! *It's* coming! Run for your lives! Nobody move! Look out! *It's* coming!"

"What's coming?" people asked. "What is *it?*"

But he wouldn't answer and he wouldn't stop. When we blocked the aisle, he jumped up and ran along the tops of the seats. "It's coming! Run! Don't make a move! Look out! Save yourselves! Eeeeyah!"

We were all pretty excited and scared. I'd call it a semi-panic. "What is *it?*" I yelled. "Hey, tell us!"

The conductor stuck his head out of the window, pulled it back in and looked at me. I could say he was the color of milk, but there's chocolate milk and buttermilk. I could say chalk, but there are all colors of chalk. There is even dirty snow and brown sugar and off-white. . . . So I'll just say he was pale and let it go. He was pale.

"Oh my gosh, it's coming, and it hates subway trains," he said. He stuck his head out again. "It eats subway trains for breakfast! Help!"

Everybody crowded around him. What is it? What does it look like? How can we save ourselves?

He was still looking out the window. "It's coming closer!"

"Describe it!"

He did. "It flies . . . and it tunnels underground. It has pockets."

"What?"

"Pockets. One hundred and seventeen pockets. And it loves to dance."

"Is that important?"

He just went on. "It can swim too, and it carries a pocket watch."

"Which pocket?" I said, but nobody paid attention.

"It's hungry and it has a can opener and a temper—it's part Scotch and part Bolivian and it's coming! Lock the windows! Everybody run! Don't make a sound! Help!"

Wow. You can guess what was going through my head at that moment. I was picturing *it* . . . that big hungry Scotch-Bolivian giant of a . . . Well, why don't you picture it yourself? Picture it in your mind, let your imagination put it together, using any colors, shapes, ideas you want. Then take a peek into your mind and see what *it* looks like. Draw it on the next page. . . .

The passengers were crunched in a bunch, all yelling at once. All I could hear were pieces of sentences: What is . . . How can . . . Please tell . . . Stop the . . . Save . . . Help . . . Then the conductor pulled his head back in, and he wasn't pale. I mean, I would say he was normal flesh color, but there are a lot of different colors for flesh and faces, right? So I'll just say he wasn't pale and let it go. He wasn't pale. And he was smiling!

"Relax," he said.
"Is it gone?"
"There isn't any it. I made it up."
"Why? Why!"

The conductor smiled and took out his pocket watch. "So you would all forget about Pittsburgh, and we could get here. And it worked." He looked at his watch. "Right on time. Pittsburgh!" And he went down the aisle calling out the word as though nothing had happened. "Pittsburgh!" Come to think of it, nothing much had happened, except in my head (Abridged from DiPego, 1970, pp. 1–13).

Quite a ride, wasn't it?

Throughout *Saturday Subway Ride,* exercises are interspersed at appropriate points. For example, in the excerpts above the student is asked to write a song, and later, to draw a picture of "it." Elsewhere in *Saturday Subway Ride,* the student is exposed to a fanciful dialogue, play script, or "tall tale," and then is asked to write his own dialogue, play script, or tall tales. After a vivid war scene, students are asked to discuss, as a class, the meaning and significance of the scene. While the workbook has not yet been formally evaluated, it seems likely that its readers will become more receptive to creative ideas, more oriented toward using their own creative capabilities, and will have a better understanding of methods for producing new idea-combinations. Especially, they should experience a keen feel for what "being creative" is like.

"WRITE? RIGHT!" A PROGRAM
FOR MORE CREATIVE WRITING [2]

Susan Houtman has incorporated principles of creative thinking and problem solving into a creative writing workbook entitled *Write? Right!* (Houtman, 1970a) prepared for upper-elementary students. In addition, by serving as a supplement to the more traditional grammar-oriented approach to language arts, the workbook provides a relevant context —writing itself—in which the rules of grammar may be learned.

As with *Thinking Creatively* and *Saturday Subway Ride,* Houtman designed *Write? Right!* to be enjoyed, to make learning fun, and to open the student to his own creative potential. A young writer's first attempts should be an adventure, a challenging game, not a chore. The student's overall incentive to write creatively may be seen as circular: something exciting can make him want to write or otherwise express himself more fully; rewarding his attempts should increase his confidence in his capability; and greater self-confidence should further increase his motivation to imagine and to express.

Hoping to shape such intrinsic motivation, *Write? Right!* exposes the student to a large variety of writing and thinking experiences—for example, short stories, advertising campaigns, rhymed and unrhymed poetry, inventing descriptive character names (e.g., "Bertha Bulgover") and story titles, songs, interviews, word games, cartoon captions, mystery plots, and letters. The student is urged, not forced, to try all exercises, no matter how new or strange they may seem (such as composing a letter his dog might write). The teacher is asked not to grade students at this point. Rather, he or she should encourage and smilingly accept whatever is produced. Students thus learn to trust their own written feelings and thoughts first, and only later worry about writing them well.

Along the way, students are shown some idea-generating techniques adapted to creative writing to help them over the "but-I-can't-think-of-anything-to-say" barrier. Some of these techniques, such as the checklist and the part-changing methods, have been described. Other strategies are a bit more unusual, such as "learning to think like a rock" or using the "old iceberg approach to ideas," in which a student learns to find unusual ideas by putting aside the first rush of ideas (the part of the iceberg everyone sees) and searching for ideas that are hidden beneath the surface. These techniques reinforce such creative attitudes as "being different is not the same as being right or being wrong"; that is, originality is a virtue.

Throughout *Write? Right!* the student is urged to visualize and verbalize in his own way. Many of the exercises are structured so that students may pick a literary, artistic, or sometimes even a musical or scientific mode of self-expression. Furthermore, the *Teachers' Manual* (Houtman, 1970b) provides several variations for approaching and handling the exercises, as well as for extending the creative thinking and writing principles in each chapter.

Conclusions

All three of our Wisconsin programs assume creative behavior is at least partly voluntary and therefore trainable. The programs thus try to amplify creative potential by fostering attitudes (including awarenesses, sets, predispositions) conducive to imaginative behavior, and by teaching some techniques for actually producing new idea-combinations.

It would be nice to report conclusively that all three sets

of materials are 100 percent effective; however, two of the three are as yet untested. The initial workbook, *Thinking Creatively* (Davis and Houtman, 1968), boasts one highly successful field study with a middle-class, seventh-grade group, and one more modest outcome in some inner-city classrooms (Davis, 1970a, 1970b; Davis, Houtman, Warren, Roweton, Mari, and Belcher, 1970). However, we are quite confident that the creativity concepts in these materials are sound and truly can add some creative flexibility to students' long-term thinking and problem-solving growth.

NOTES

1. Novel idea-combinations also can occur as involuntary, non-stoppable mental events—try *not* to think of a red-and-white-striped gorilla, and don't let him eat that blue banana.
2. This section was prepared in collaboration with Susan E. Houtman.

◄ CHAPTER 11 ►

MORE IMAGINATION AND PROBLEM SOLVING IN THE CLASSROOM

IN RECENT YEARS, several scholars have produced training materials directed at putting more imagination and problem solving into the classroom. In this chapter we will examine the thinking behind three structured approaches, the Myers and Torrance Idea Books (1964, 1965a, 1965b, 1966a, 1966b, 1968), the Covington, Crutchfield, and Davies (1966) *Productive Thinking Program,* and the Purdue audio tapes and materials, *The American Pioneers* (Feldhusen, Treffinger, and Bahlke, 1970; WBAA, 1966).[1] The respective formats of these three innovations reflect the viewpoints and strategies of the particular group of authors and therefore the materials are quite different from one another. However, all emphasize attitudes consonant with flexible thinking, and all three seek to strengthen through practice some relevant problem-solving skills and abilities.

Myers and Torrance Idea Books: Exercising Creative Abilities

It seems reasonable to suppose that some mental abilities, like athletic skills, might be improved with practice. As to the recommended specific form of this practice, Guilford (1962) suggested that particular creative abilities might be improved by giving exercises similar to tests that measure those abilities. For example, if the number of responses to such tests as "List things which are *solid, white,* and *edible*" or "Name synonyms for the word *dark*" is a good indicator of the *fluency* ability, then the use of these or similar tests as exercises might strengthen this ability.

In addition to sharpening basic abilities, creative thinking exercises also may be used to teach attitudes appropriate for imaginative behavior: for example, by asking for and rewarding "wild" idea-combinations.

The six Idea Books prepared by Myers and Torrance (1964, 1965a, 1965b, 1966a, 1966b, 1968) employ clever and well-illustrated exercises to accomplish both purposes: to strengthen simple and complex thinking abilities and to foster attitudes conducive to imagination.

My own list (Davis, 1969) of the abilities practiced in the Idea Books includes remembering, free-associating, sensing problems, perceiving relationships, imagining and elaborating upon wild ideas, predicting or making up consequences of unusual events, filling in information gaps (verbal and visual), pretending, and being more aware of sensory experiences. Further, some Idea Books provide practice in using descriptive adjectives; finding unusual uses; and making up story plots, puzzles, punch lines, mysteries, and even more exercises.

To illustrate, the youngest mind-stretching Idea Book, *Can You Imagine?* asks, "What could happen if cats could

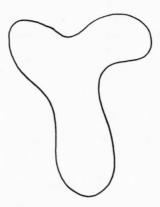

Figure 11–1 · Nonsense stimulus figure similar to one used by Myers and Torrance (1965a) to elicit imaginative responses.

bark when they wanted to?" and "When is the sky?" Given a blob similar to the one in Figure 11–1, the youngsters are asked, "Can it talk or sing? What does it sound like? Can it walk or jump? How does it move?" Another imagination-prodder asks, "If you were on the moon, would you rather be a deer or a frog? Why? Show how you would be a frog or deer by dancing like one, by singing about what you might do if you were one, by writing about what you would do if you were one, by drawing what you think you would do—right here."

Working on both memory and sensory awareness, exercises in *For Those Who Wonder* ask, "Do you remember when you smelled [tasted, saw, touched] something very nice for the first time? What was it? What did you do? How did you feel when you smelled [tasted, saw, touched] it?"

More advanced Idea Books add creative writing skills to stimulating imagination. For example, *Plots, Puzzles and Ploys* asks for similes ("Frisky as ———"), short stories, speeches, and personal letters (written by your bicycle), and even requests ideas for an advertising campaign to sell leftover Beatle wigs.

Crutchfield and Covington (1965) once suggested that to teach skills of creative thinking, one must use creative tactics. Undoubtedly, Myers and Torrance have done this.

The Purdue Creativity Program

Prepared for elementary grades three through six, *Creative Thinking: The American Pioneers* (Feldhusen, Treffinger, and Bahlke, 1970; WBAA, 1966) is a set of twenty-eight audio tapes, with three or four printed exercises to accompany each tape.[2] Delightful scripts for the recorded materials were prepared in collaboration with staff members of WBAA's "School of the Air" (Lafayette, Indiana) and were narrated by a professional radio announcer. The tapes include appropriate sound effects: for example, hoof beats for a Pony Express saga along with suitable background music.

Three components of each tape reflect the central goals of the materials. First, a brief presentation at the outset of each tape verbally describes a few attitudinal principles of creative thinking. For example, the children hear about the sociohistorical importance of effective thinking, and they are instructed to be open, flexible, and receptive to new ideas. The presentation further explains that creative thinkers produce ideas (sometimes humorous ones), solve problems, and create original products by putting ideas together in new ways.

Following the opening presentation of creative problem-solving principles, the students hear a story depicting some historical event. Most of the stories contain obvious examples of problem solving, creativity, discovery, or some other form of adventurous effective thinking. Four categories of stories include *Early Explorers* (describing the lives of, for example, Balboa, Columbus, Cortez, Magellan, and Lewis and Clark), *Frontiers in Transportation and Communication* (about Samuel Morse and the Telegraph, The Pony Express, Laying the First Cable, Wright Brothers, Henry Ford, Television, and Telstar), *Fighters for Freedom* (Washington, Lincoln, Simon Bolivar, Theodore Roosevelt, and Eisenhower), and *Frontiers in Health and Science* (for example, Fleming and Penicillin, Conquering Yellow Fever, Salk, Sabin and Polio, Goddard and the Rocket, and John Glenn and Gus Grissom).

Following the creative thinking principles and the adventure, the third training component is the exercises themselves. According to developers Feldhusen, Treffinger, and Bahlke, the exercises seek to strengthen the thinking abilities of *originality, flexibility, fluency,* and *elaboration.* To illustrate, one exercise taken from *Henry Ford* reads:

Suppose that Henry Ford had not invented an automobile, and suppose that we had no motors of any kind today. How many things would be different? What things would you not be able to do? List all the things you can think of that you would not be able to do.

And from *Henry Hudson:*

Hudson was bound and placed aboard the small boat along with his son and several sick members of the crew. The rest of the crew sailed away. What happened to Hudson and his men we do not know. They were never seen again. What do you think happened? Suppose that Hudson, his son, and the sick crew members did not die immediately but lived for a long time. Tell the story of their lives from the time they were set adrift in the small boat.

One exercise from *Jonas Salk* stresses scientific caution while stimulating idea-production:

Finally, in 1953, Jonas Salk thought he had succeeded. He had injected mice with a vaccine and then had given them a killing dose of polio virus. But now the mice lived. Their legs did not become paralyzed. Dr. Salk did not celebrate. He was a good scientist and good scientists are cautious. Why do you think Dr. Salk did not celebrate? Why was he cautious? List all the good reasons you can think of.

Evaluations of the Purdue Creativity Program by Feldhusen and others (Bahlke, 1967, 1969; Feldhusen, Bahlke, and Treffinger, 1969; Robinson, 1969) showed the materials to be effective in improving creative thinking skills measured by the Torrance Tests of Creative Thinking (TTCT; Torrance, 1966). A study evaluating the effects of the three separate components of the training experience—the creative problem-solving principles, the story, or the exercises—showed overall that the printed exercises seemed especially responsible for most of the improvement in TTCT scores.[3]

The Productive Thinking Program

The *Productive Thinking Program* (Covington, Crutchfield, and Davies, 1966) is a set of sixteen "semi-programmed" lessons written primarily for fifth-grade and sixth-grade students.[4] The materials take the form of rather intriguing detective stories. Each mystery nicely serves the dual function of maintaining keen interest, especially since the student himself participates in cracking each case, while teaching meaningful principles of problem-solving and creative behavior.

A very clever story line includes two cartoon characters of the students' age, Jim and Lila, plus Jim's Uncle John, a part-time incognito sleuth known as Mr. Search. While leading Jim, Lila, and the student-participant through each enigma, Mr. Search stresses making good use of information (for example, in idea-evaluation), becoming sensitive to important problem clues (literally), generating high quality ideas, asking important questions, and being "planful" when solving a problem. Uncle John further encourages the development of such attitudes as openmindedness (which includes looking for unusual ideas, not jumping to conclusions, and accepting mistakes), self-confidence in the ability to improve one's thinking skills, and (hopefully) a general liking for thinking tasks.

The Berkeley group has summarized a number of studies demonstrating positive effects of their materials upon such dependent measures as Covington's (1966) attitude inventory and more detective mysteries and other convergent problems (such as Duncker's X-ray problem). They also describe transfer effects to such tasks as *picture completion* problems, *unusual uses, unusual consequences,* and *product improvement* tasks (Covington and Crutchfield, 1965; Crutchfield and Covington, 1965; Olton and Crutchfield, 1969; Olton, Wardrop, Covington, Goodwin, Crutchfield, Klausmier, and Ronda, 1967). One other research report, however, was less favorable (Treffinger and Ripple, 1969).

Conclusions

The creators of these materials, Myers and Torrance, Feldhusen and his group, and the Berkeley group of Covington, Crutchfield, and their colleagues, share some strong beliefs. First of all, they realistically are convinced that conventional education, even though there are many different and some progressive conventions, does not adequately develop thinking and creative problem-solving skills. They further assume that enough *is* known about problem solving and thinking to develop theoretically and intuitively sound materials and teaching strategies which can help correct thinkless education. As described earlier in this volume, and especially in the preceding chapter, we *can* teach students some appropriate *attitudes,* some idea-finding and systematic problem-solving *techniques,* and following the Myers and Torrance strategy, we can strengthen through practice and experience some basic *abilities* underlying creative behavior. With our principles and objectives fairly well-defined, systematic training in problem solving and creativity should develop students' thinking skills just as training in music and mathematics will improve musical and mathematical skills.

NOTES

1. Several excellent materials are not covered in this chapter. In language arts, especially, the educator may be interested in the Ginn 1970 *Reading 360* series (levels 1–13) or the Harper and Row *New Directions in English* (grades 1–8). Also, Smith and his colleagues have published a series dealing with creative teaching of science, language arts, creative arts, mathematics, and social studies in the elementary school (Allyn & Bacon, Inc.).

    While the Idea Books and the *Productive Thinking Program* are commercially available, *American Pioneers* seems readily available only to member stations of the National Association of Educational Broadcasters.
2. Another set of audio materials is the Cunnington and Torrance *Imagi/Craft* records, dramatizations and creative thinking exercises intended to supplement the elementary curriculum in science, history, geography, and language arts.
3. Aware of the unfortunate white bias in American history education, Feldhusen and Treffinger at the time of this writing are preparing similar materials describing Black American pioneers.
4. While the materials do not appear in a traditional, frame-by-frame programmed form, the booklets do elicit and reinforce students' responses throughout.

◀ PART IV ▶

Summary

◄ CHAPTER 12 ►

REVIEW AND COMMENT

THE GOAL of this short book was to provide a state-of-the-art overview of research, theory, principles, and procedures of creative problem solving. While approached differently by explanation-minded psychologists, on one hand, and the application-oriented educators and business people on the other, problem solving continues to intrigue thinkers in all of these areas. And indeed it should, for certainly creative flexibility is a state of mind, a philosophy of life allowing one to recognize the blinders of habit and tradition and enjoy the excitement of considering new ideas and creative innovations.

In the opening chapters, we reviewed—perhaps unnecessarily—the critical significance of problem solving and creativity in our individual lives and in the history of civilization. As further groundwork, our definitions of a *problem* and a *solution,* and our discussions of problem-solving *stages* and the negative effects of *habit* and *conformity,* led eventually to a taxonomy of problem-solving tasks, based upon the nature of activities elicited by a problem stimulus. All the simplifying definitions and assumptions, and the taxonomy, were intended to call attention to important commonalities (and differences) across a rather astounding variety of problem encounters.

Turning to psychological accounts of human problem

solving, we visited Köhler's primate station in the Canary Islands. Köhler's studies with chimps, children, and chickens demonstrated the Gestalt concept of "insight" or "cognitive restructuring." We also looked briefly at the Luchins' water-jar experiments and their emphasis on *rigidity* of behavior growing out of strong, interfering habits. Such rigidity would be overcome by Maier's concept of problem-solving *direction* or Sheerer's notion of *recentering,* a matter of shifting one's approach to (what else?) the correct direction. Bruner, representing neocognitive psychology, explained how humans uniquely use *higher-order coding*—categorizing, attaching probabilities to events, applying principles or theories—to go beyond the given information, intelligently solving problems and making decisions.

If Gestalt and cognitive explanations of problem solving are too vague, *S-R* ideas are too simple, too mechanistic. Transfer, trial-and-error, response-hierarchies, even the notion of implicit (mediating) associations, *do* indeed simplify the dynamics of much observable problem solving. However, too much is ignored. Our beautifully human problem solving always is conscious, deliberate, directed, and strategic; and such thinking just cannot be totally described in the language of conditioned responses.

The efforts of the computer-simulation specialist to duplicate human problem solving require him to detail a complete sequence of humanlike information processes. His programming approach thus resembles that of the cognitive-Gestaltist in their mutual concern with phenomenological, cognitive-level thinking processes. To date, the relatively new and obviously specialized area of computer-simulation of problem solving, part of the larger field of artificial intelligence work, is primarily instructive in calling attention to previously ignored (or unrecognized) details of sequentially organized information-seeking and information-processing strategies. Such strategies invariably include *heuristics,* humanlike means-end checks that allow the computer—and, by analogy, the human problem-solver—to select routes not by trial-and-error, but by conscious goal-oriented shortcuts.

While the psychologist and his computer sciences colleagues busily study basic variables and processes in problem solving and thinking, the educational and industrial folk try, with some success, to teach principles and procedures for finding creative problem solutions. Our concepts of creative *attitudes* and creative thinking *techniques* summarize in two words the main content of most professional-level problem-solving courses and programs. Trainees, usually engineers and executives, but often including all levels of workers and managers, become more aware of the importance of innovation, more receptive to strange ideas, and they learn some deliberate idea-producing techniques to supplement their native ingenuity. Brainstorming, attribute listing, morphological synthesis, checklisting, and the metaphor-based synectics and bionics methods, all may increase the flexibility of the individual problem-solver. If we wished, we could say that such techniques provide the "recentering" or new approaches described by Sheerer, the "direction" recommended by Maier, or that they lead to Köhler's "insights" or serve as Bruner's "higher-order codes."

In the classroom, the Dewey-Suchman-Shulman concept of inquiry activity reminds us that problems in the real world are not neatly defined and presented to the individual, as they are in the experimental laboratory, nor is the point of solution always clear. Rather, the human encounter involves sensing the problem, a disequilibrating experience; formulating the problem; searching for relevant information; and resolving the felt difficulty to the individual's personal satisfaction. Researchers Shulman, Suchman, Butts, and others have discovered that materials and gamelike role-playing methods give valuable practice in the type of scientific problem solving described by the four steps.

Other programs and materials in education approach the teaching of problem solving and creativity quite differently. Our own materials have tried to translate creative problem-solving principles—attitudes and techniques—into a form suitable for upper-elementary and junior high school students. Working with fifth-grade and sixth-grade students, Covington and his Berkeley colleagues also teach attitudes conducive to effective, flexible thinking, along with a general problem-solving strategy resembling the four inquiry steps. The multifaceted workbooks of Myers and Torrance seek to teach attitudes and techniques related to imagination, but most of all try to strengthen through practice some important basic problem-solving *abilities*. Finally, through presenting a radio show depicting creative thinking by *American Pioneers,* the Purdue group led by Feldhusen has registered fine results in increasing fluency, flexibility, and originality of upper-elementary-school children.

The study of problem solving and creativity obviously has taken many forms, as it should, since problem solving and creativity themselves appear in every conceivable circumstance. Furthermore, no theoretical or empirical approach has a corner on truth. Rather, the approaches supplement each other, particularly, and I suppose, indirectly, by pointing out important dimensions of creative problem solving which other viewpoints have ignored. No doubt psychology, industry, and education have much to say to each other on these topics of mutual interest. I hope this volume is a small step in improving this communication.

◄ APPENDIX A ►

TESTS AND MEASURES OF CREATIVITY

THE FOLLOWING instruments are listed for the interest, information, and potential use of the reader. Listing a particular instrument in no way implies endorsement. The many shapes and forms of creative abilities and creative problem-solving behaviors, along with drastic differences among the tests themselves, do not allow one to easily judge the worth of each of dozens of measuring devices. To note just a few more complications, some tests measure potential for creative behavior (a capacity which may never be used fully) by tapping attitudes, self-concepts, and motivations, or else by sampling people's ability to produce original ideas; other instruments measure actual creative productivity by assessing creative activities engaged in or awards won. Often, scores from different tests will correlate more highly with I.Q. than with other creativity test scores, suggesting that the "creativity" test functions only as a (perhaps unreliable) intelligence measure (Wallach and Kogan, 1965). Also, the administrator's instructions, or one's self-instructions, may severely alter a creativity score (Manske and Davis, 1968). We must leave it for the reader, then, to determine as carefully as he can if a given instrument suits his particular needs.

The measuring devices are divided into four categories: (1) Commercially Available Tests, (2) Commercially Available Interest, Attitude, and Biographical Inventories, (3) Non-Commercially Available Tests, and (4) Non-Commercially Available Attitude Surveys. Although the instruments listed in the third and fourth categories may or may not appear in their entirety in published articles, at least one reference is cited where more complete information (including purposes, scoring details, factor loadings, etc.) may be located. The authors of the reference should not be taken as the originators of the test; while Guilford or Thurstone usually are responsible, it would be a truly difficult task to trace the inventor of each and every test of creative ability.

Commercially Available Tests

Guilford Tests. A number of tests developed by Guilford and his colleagues are commercially available: for example, *Guilford-Zimmerman General Reasoning, Logical Reasoning, Word Fluency, Expressional Fluency, Ideational Fluency, Associational Fluency, Alternate Uses, Consequences,* and *Decorations.* Beverly Hills, Calif.: Sheridan Supply Company.

Remote Associates Test, by S. A. Mednick. In each of thirty items, S is given three words and is asked to find a fourth word related to all three. Boston: Houghton Mifflin Company, 1967.

Sounds and Images, by B. F. Cunnington and E. P. Torrance. Familiar and obscure sounds on 33⅓ rpm record elicit associations that may be scored for originality.

Torrance Tests of Creative Thinking. Verbal Forms A and B contain seven subtests, requiring S to (1) ask questions about an unusual picture, (2) guess causes of the action in the picture, (3) guess consequences of the action in the picture, (4) think of ideas for improving a stuffed toy monkey or elephant, (5) list unusual uses for cardboard boxes or tin cans, (6) ask unusual questions about cardboard boxes or tin cans, and (7) predict consequences of an improbable event. Figural Forms A and B contain three subtests, all of which require S to draw pictures that elaborate upon (1) a single brightly colored form, (2) 10 incomplete line drawings, and (3) 36 identical circles (or pairs of parallel lines). Princeton, N.J.: Personnel Press, Inc., 1966. See also Torrance (1962, 1965).

Commercially Available Interest, Attitude, and Biographical Inventories

Adjective Check List, by H. G. Gough. Self-rating checklist comprised of 300 adjectives. Palo Alto, California: Consulting Psychologist Press, 1952. See Domino (1970) for creativity scoring key.

Alpha Biographical Inventory. 300 multiple-choice items assess potential for scientific creativity in high school students. Greensboro, N.C.: Institute for Behavioral Research in Creativity, 1966.

Biographical Inventory, by C. E. Schaefer. 125 multiple-choice items assess biographical information and creativity-related activities in areas of Family History, Educational History, Avocational Activities, and Miscellaneous. San Diego: Educational and Industrial Testing Service, 1970.

Opinion, Attitude, and Interest Survey (OAIS). Questionnaire keyed for several traits, including creative potential. OAIS Testing Program, Box 380, Ann Arbor, Michigan.

Zimmerman-Guilford Interest Inventory. Fifteen-item true-false scale indicating interest in various creative activities. Beverly Hills, Calif.: Sheridan Supply Company, 1963. See also Taft and Gilchrist (1970).

Non-Commercially Available Tests [1]

Anagrams (multiple solution). S makes as many words as possible from the given word or letters (Guilford and Merrifield, 1960; Thurstone, 1944).

Apparatus Test. S suggests two improvements for each of several appliances (Guilford et al., 1951, 1952).

Associations. S gives word related to both of two given words (Guilford, 1967).

Camouflaged Words. S finds the name of a sport or game concealed in a sentence (Guilford and Merrifield, 1960).

Cartoons. S writes one punch line for each of several given cartoons (Guilford, 1967; Christensen and Guilford, 1963).

Common Situations. S lists problems suggested by everyday situations (Guilford et al., 1951, 1952).

Concept Synthesis. S combines two ideas to suggest new idea (Guilford et al., 1951, 1952).

Consequences. S thinks of consequences of unlikely event (for example, sleep no longer needed). Other names: *Just Suppose, Unusual Consequences, Guessing Consequences, Novel Situations.* See Torrance (1962).

Controlled Associations. S writes several synonyms, or words similar in meaning, to each given word (Guilford et al., 1951, 1952).

Decorations. Given outlines of common objects (for example, furniture, clothing) S adds decorations (Guilford, 1967).

Designs. Given simple line elements (angle, circle, bow, dot), S makes designs which he might see in wallpaper, fabric, etc. (Guilford, 1967).

Drawing Completion Test. S elaborates on simple figures (Barron, 1958).

Fables Test. Four fables have missing last lines. S supplies a moralistic, humorous, and sad ending for each (Getzels and Jackson, 1962).

Figure Preference Test. S expresses preference for abstract drawings differing mainly on complexity-simplicity dimension (Barron, 1958). Author's note: Artistic tastes have changed substantially in the past decade. At present, this test likely is completely invalid.

Figure Production. Given one or two lines, S makes meaningful object by adding lines (Berger, Guilford, and Christensen, 1957).

First and Last Letters. S writes words given first and last letters (French, 1951; Guilford, 1967).

Gestalt Transformation. S indicates which of five listed objects contains a part that will serve a specified purpose (Guilford and Merrifield, 1960).

Appendix A

Hidden Figures. S indicates which of several simple figures is located in more complex geometric form (Witkin, Dyk, Faterson, Goodenough, and Karp, 1962).

Hot Dog Problem. S invents new kinds of hot dogs by improving weiners, buns, or both (Davis *et al.,* 1969).

Implied Uses. S gives several secondary meanings of words (Guilford *et al.,* 1951, 1952).

Impossibilities. S lists things that are impossible (Guilford *et al.,* 1951, 1952).

Independence of Judgement Test. High-scoring Ss resist yielding to group pressure in responding to opinion questionnaire (Barron, 1958).

Instances. S generates examples of class concept (for example, round things, things that make a noise). Other names: *Object Naming, Thing Categories* (French, 1951; Wallach and Kogan, 1965).

Inventive Opposites. S is given words for which he produces two words each, opposite in meaning to the given words (Guilford, 1967; Thurstone, 1944).

Letter Star Test. Given letter-asterisk-letter-asterisk (for example, Y*N*), S constructs as many four-word sentences as he can, using the letters as initial letters of two words (Carroll, 1941).

Line Meanings. S thinks of meanings or interpretations of abstract designs drawn with single line (Wallach and Kogan, 1965).

Make a Figure Test. Given two or three line segments, S makes series of figures using only these lines (Guilford, 1967).

Make a Mark Test. S makes as many simple line figures as he can, within specifications (for example, making open figures using dotted lines, closed figures with curved lines) (Guilford, 1967).

Make Up Problems Test. Given complex paragraph containing numerical information pertaining to, for example, buying a house or building a swimming pool, S makes up as many mathematical problems as he can (Getzels and Jackson, 1962).

Making Objects. S is asked to make a particular object, using any combination of four simple figures (Guilford, 1967).

Matchstick Problems. Given a pattern made of matchsticks, S is asked to move or remove a certain number of matchsticks to create another specified pattern (Cline, Richards, and Needham, 1963; Guilford, 1967; Katona, 1940). In one variation, S is asked to construct four equilaterial triangles from six matches, requiring a three-dimensional solution.

Mechanical Principles. S applies simple mechanical principles to solve problems (Guilford *et al.*, 1951, 1952).

Mosaics. Given colored pasteboard squares, S designs own mosaic pattern (Barron, 1958).

Number Associations. S lists associations for given numbers (Guilford *et al.*, 1951, 1952).

Object Synthesis. S names object that could be made by combining two given objects (Guilford and Merrifield, 1960).

Picture Gestalt. S indicates which object in a photograph will serve a specified purpose (Guilford and Merrifield, 1960).

Planning Elaboration. S fills in as many details as necessary to make a briefly outlined activity work (Guilford and Merrifield, 1960).

Plot Titles. S produces titles for short story plots (Guilford, 1967; Johnson and Zerbolio, 1964).

Possible Jobs. Given a meaningful pictorial design (for example, a light bulb), S thinks of occupations the picture might represent (Guilford, 1967).

Practical Judgment. S finds best solution to given practical problem (Guilford *et al.*, 1951, 1952).

Prefixes. S writes words beginning with a specified prefix (French, 1951; Guilford and Merrifield, 1960).

Product Improvement. S thinks of improvements for given object (Torrance, 1962).

Production of Figural Effects. Given one or two lines, S adds lines *without* making a meaningful figure (Guilford, 1967).

Quick Responses. With five seconds per item, S gives free associations to each of fifty stimulus words (Guilford, 1967).

Rhymes. S lists words rhyming with specified word (Guilford, 1967).

Riddles. S produces two solutions for each of several riddles; one solution is to be clever (Guilford, 1967).

Seeing Deficiencies. S lists ways in which a described plan is faulty (Guilford and Merrifield, 1960).

Seeing Problems. S lists problems that might arise in connection with common objects (Guilford and Merrifield, 1960).

Sentence Fluency. S restates given idea in different ways (Taylor, 1947).

Similarities. S thinks of similarities between two objects (for example, carrot and potato) (Wallach and Kogan, 1965).

Simile Insertions. Given the beginning of a simile (e.g., his smile was wide as a _____), S fills in as many alternate words as he can (Guilford, 1967).

Simile Interpretations. S completes sentence which states analogous idea (Guilford and Merrifield, 1960).

Simile Interpretation Test. S provides different interpretations of a given simile (Christensen and Guilford, 1963).

Sketches Test. Given a series of simple basic figures (for example, circles), S creates as many real objects as he can with a minimum addition of lines (Guilford, 1967).

Social Institutions. S suggests two improvements for each of several social institutions (for example, marriage, sales tax) (Guilford *et al.*, 1951, 1952).

Suffixes. S writes words ending in specified suffix (for example, -tion) (French, 1951; Guilford, 1967).

Symbol-Equivalence Test. S free-associates to "stimulus image" (for example, empty bookcase, sound of foghorn) (Barron, 1958).

Symbol Production. S produces symbols to represent given activities and objects (Guilford and Merrifield, 1960).

Topics. S writes as many ideas (or words) as possible about a given topic (French, 1951; Guilford and Merrifield, 1960).

Topics. S writes on a given subject. Other name: *Theme* (Guilford, 1967).

Tourist Problem. S writes proposal (essay) for getting more European visitors to the United States (Hyman, 1964).

Triangle Puzzle. S makes four equilateral traingles with six matches.

Two-Word Combinations. S produces as many two-word sentences as he can. Variations include specifying first letters for the two words, using four-word sentences, using four-word sentences with first letter of each word specified (Guilford, 1967).

Unfinished Stories Test. S completes story (Taylor, 1947).

Unusual Details. S indicates anomalous features of pictures (Guilford et al., 1951, 1952).

Unusual Methods. Given a problem (for example, bored employees), S thinks of two *unusual* solutions (Guilford, 1967).

Unusual Uses. S lists uses for common object (for example, a brick, wire coat hanger). Other names: *Alternate Uses, Utility Test, Brick Problem, Hanger Problem.* Variation: May limit S to six responses per object (Guilford, 1967; Torrance, 1962).

Word Arrangement. S writes sentences containing four given words (Christensen and Guilford, 1963).

Word Associations. S writes as many meanings as he can for each of twenty-five words with multiple meanings (Getzels and Jackson, 1962).

Untitled Test. Child writes about himself as he might be twenty-five years in the future (Flescher, 1963).

Untitled Test. S writes words with first letter given (Guilford, 1967).

Untitled Test. S writes words with one (or two, or three) letters specified (Guilford, 1967).

Untitled Test. Given ten various shaped blocks, S designs home furniture (Welch, 1946).

Untitled Test. S writes story using twenty given words (Welch, 1946).

Untitled Test. S forms alphabet letters from a few given lines (Welch, 1946).

Untitled Test. S composes sentences from ten randomly presented words (Welch, 1946).

Non-Commercially Available Attitude Surveys

Childhood Attitude Inventory for Problem Solving, by R. S. Crutchfield and M. V. Covington. Assesses attitudes related to creative and problem-solving activities (Covington, 1966).

Empathy Scale. Ten items measure capacity to empathize with story characters or real people (Elms, 1966).

Experiences Questionnaire. Seventy-nine yes-no items tap ac-

Appendix A

tual and imagined experiences which may relate to tendency for creativity (Taft, 1969; Taft and Gilchrist, 1970).

How Do You Think? Twenty items, rated on nine-point scales (never true—always true) assess attitudes, motivations, and self-perceptions related to creativity and problem-solving behavior (Davis, Houtman, Warren, and Roweton, 1969).

Pennsylvania Assessment of Creative Tendency (PACT). Forty-five self-rating items measure attitudes and interests related to creativity (Rookey, 1967).

Personal-Social Motivation Inventory. S endorses whichever of 30 items apply to him; measures creative interests and motivations (Torrance, 1971).

Preconscious Activity Scale. Thirty-eight items reflect respondent's attitudes toward engaging in creative activities (Holland and Baird, 1968).

Thinking Interest Inventory. Assesses attitudes and motivations pertaining to creative behavior (Merrifield, unpublished test).

What Kind of Person Are You? Fifty two-choice items reflect attitudes contributing to creative potential (Torrance, 1971).

NOTE

1. Some individual tests listed in this section are subtests of commercial batteries listed above. Also, a few tests, or their variations, will be listed in Appendix B as laboratory problem-solving tasks.

◄ APPENDIX B ►

LABORATORY PROBLEM-SOLVING TASKS

IN DEVISING a problem-solving situation for laboratory research, the only apparent limitation is one's imagination. Despite the fact that about half of recent research in problem solving has evolved about the anagram task, the remaining problems are very heterogeneous, to say the least, in the nature of materials and in the problem-solving activities elicited. The person searching for a laboratory task should find something in this section which fits his needs. If not, with a little more effort he may create a unique task of his own without committing a significant impropriety.

Most of the task names are widely agreed upon. In a few cases, I confess, I have taken the liberty of assigning a name which to me seemed eminently suitable. For example, a task requiring S to decipher sentences I have called the "deciphering problem." Still, there are a number of "unnamed tasks." Very familiar problems (such as arithmetic problems or jigsaw puzzles) are just named, without explanation. For nearly all tasks, a reference is provided.

Regarding omissions, the large variety of tasks called *concept learning* problems (card-sorting and object-sorting tasks, along with a variety of verbal and other concept tasks) are mostly ignored; the reader interested specifically in laboratory concept learning should see Bourne (1966). Also, many tasks used to

Appendix B

study interpersonal relations (for example, leadership emergence, effects of communication barriers, etc.) in small groups, have been omitted. For these, see Zagona, Willis, and MacKinnon (1966) and Maier (1970). Finally, I also should mention that it is common for the researcher to use many of the creativity tests listed in Appendix A as laboratory problem-solving tasks.

Algebra Word Problems. Ss solve verbally presented algebra ("story") problems (Paige and Simon, 1966).

Alphabet Maze. Given a matrix with a letter in each cell, S tries to "move" from top-right to bottom-left corner by spelling words (Cowen, Wiener, and Hess, 1953).

Alternation Problem. S learns to predict regularity in single (ABAB), double (AABB), triple, or quadruple stimulus alternation series (Linker and Ross, 1963).

Anagrams. S makes meaningful word out of scrambled letters. See also Multiple-Solution Anagrams (Johnson, 1966).

Analogy Problems. S solves verbal or numerical problems of the form "A is to B as C is to _____?" (Gentile, Kessler, and Gentile, 1969; Johnson, 1962).

Ball Problem. Same as Pea Transfer Problem.

Bent-Wire (or *Bent-Nail*) *Puzzles.* S searches for secret moves to untangle two strategically twisted nails (Blumenfeld, 1964).

Candle (or *Box*) *Problem.* Given pasteboard box, matches, candle, and thumb tacks, S devises means of mounting candle on wall. In variation, S may be blindfolded (Suedfeld, Glucksberg, and Vernon, 1967).

Card Tricks. See Katona (1940).

Chess Problems. With pieces in prepared positions, S searches for good (perhaps winning) sequence of moves (DeGroot, 1965, 1966).

Circuit Problem. Given insufficient lengths of wire, S must discover that screwdriver may be used to complete circuit (Glucksberg, 1964).

Conservation Problems. Liquid is transferred to taller, thinner jar, and children decide if the amount of liquid remains the same (Frank, 1966).

Crossword Puzzles.

Cylinder in the Can Problem. S thinks of ways to remove a solid wooden cylinder from the inside of a close-fitting can (Ray, 1966).

Deciphering Problem. S deciphers ten sentences; each sentence contains eighteen letters transposed according to a systematic rule (Wittrock, 1963).

Detour (or *Umweg*) *Problem.* With goal in view behind barrier, Ss (usually animals) must learn to first move away from the goal, around the barrier, and then to the goal (Köhler, 1925).

Disc Transfer Problem. See Egyptian pyramid puzzle.

Doodlebug Problem. S must overcome strong habits or "everyday beliefs" to puzzle out correct moves for imaginary bug (Lyda and Fillanbaum, 1964).

Egyptian Pyramid Puzzle. Three (or more) different-sized rings are stacked on one of three pegs. S's task is to transfer the rings to another peg, moving one ring at a time and never placing a larger ring on a smaller one. Other names: Disc Transfer Problem, Tower of Hanoi Problem, Pyramid Problem.

Farmer Problem. S tries to divide L-shaped piece of land into four parts equal in size and shape (Maier, 1970).

Four-word Problems. S selects word unrelated to other three (Judson and Cofer, 1956).

Gold-Dust Problems. Basically the same as water-jar problems, with gold dust substituted for water (Restle and Davis, 1962).

Hatrack Problem. S is asked to construct "hatrack" from an assortment of boards, a C-clamp, and other junk (Maier, 1970). Related to pendulum and two-string problems.

Hidden (or *Embedded*) *Figure Test.* S tries to locate simple figure embedded in complex design (Witkin *et al.*, 1962).

In-Basket Problem. S plays role (for example, of teacher or mayor), responding to potential problems in in-basket. See Chapter 6.

Jigsaw Puzzles.

Last Draw Problem. Ten paper clips are laid on table. Given first draw, of one or two clips, S learns strategy for getting last clip (Ray, 1966).

Letter-Series Problems. S predicts next letter in sequence organized according to some rule. Similar to number series problems.

Matching Problems. From presented stimuli, S abstracts rule allowing him to select additional stimuli according to that rule (Goldiamond, 1966).

Matchstick Problems. Given a pattern made of matchsticks, S is asked to move or remove a certain number of matchsticks to create another specified pattern (Katona, 1940).

Mathematics Problems. See Polya (1957), Kilpatrick (1969). Also see algebra word problems.

Mayor Problem. See in-basket problem.

Mined Road Problem. Given assortment of junk, Ss devise means of crossing heavily mined road (Lorge et al., 1956).

Missionaries and Cannibals Problem. Three missionaries and three cannibals must cross river in two-man boat without ever leaving more cannibals than missionaries on either side (Simon and Newell, 1961).

Multiple-Solution Anagrams. Given moderately long word or assortment of letters, S creates as many meaningful words as he can (Kaplan and Carvellas, 1969).

Non-Reversal (or *Extradimensional*) *Shift Problems.* For example, with four stimulus patterns, *black triangle, black square, white triangle,* and *white square,* S first is reinforced for responding to black figures (*color* dimension). Then without notice or interruption, reinforcement becomes contingent upon responding to triangular figures (*shape* dimension) (Kendler and Kendler, 1962). See reversal shift problems.

Number Series Problems. S predicts next number in sequence organized according to some rule (Simon and Kotovsky, 1963; Wason, 1968). Similar to letter series problems.

Oddity Problem. When exposed to three stimulus objects or patterns, two of which are identical, Ss (usually monkeys or small children) are reinforced for selecting the odd stimulus (Gollin and Shirk, 1966).

Pea (or *Ball*) *Transfer Problem.* Using provided apparatus, S invents ways to transfer peas from one container to another from a distance (Raaheim, 1963).

Pendulum Problem. S is asked to construct two pendulums

from an assortment of boards, a C-clamp, pieces of string, and other junk (Maier, 1970). Related to hatrack and two-string problems.

Prisoner Problem. S tries to trace prisoner's path through 4 by 4 matrix of prison cells, under various constraints (Maier, 1970).

Probability Learning. With just two or three specified response alternatives available (for example, buttons, knobs, verbal responses), each of which "pays off" on a fixed percentage basis, S solves the problem by always selecting the response with the highest payoff percentage (Odom, 1967).

Puzzle Box. Jail-like box used with animal problem-solving to study solution (escape) behavior (Thorndike, 1911).

Pyramid Problem. See Egyptian pyramid puzzle.

Reversal (or *Intradimensional*) *Shift Problems.* For example, with four stimulus patterns—*black triangle, black square, white triangle,* and *white square*—S first is reinforced for responding to black figures. Then without notice or interruption, the reinforcement contingency is "reversed" to white figures (Kendler and Kendler, 1962). See nonreversal problems.

Rotating Weights Problem. In a physics-type demonstration, S learns that propeller-like apparatus will be easier to start spinning if weights are placed near center, rather than at extremes. S then is asked whether hollow or solid ball (equalized in size, weight, and appearance) will roll down a slope faster (Anderson and Johnson, 1966).

Spy Problems. After memorizing a list describing which spies can talk to which other spies, S figures out how to get a message from one particular spy to another particular spy (Hayes, 1966).

Stick (or *Banana*) *Problem.* Chimpanzee learns to use stick to retrieve banana, which is out of reach overhead or outside of cage.

Square Roots Problems. In computing square roots with a desk calculator and partly falsified table, S must find errors in table (Marks, 1951).

Switch-Light Problems. S tries to produce required light pattern by manipulating switches (Davis, 1967).

Syllogistic Reasoning (or *Logic*) *Problems.* Given set of

Appendix B

propositions—for example, all A are B, some B are C—S chooses correct implication (Frandsen and Holder, 1969; Henle, 1962; Moore and Anderson, 1954).

Tab Item Test. A printed test of diagnostic, trouble-shooting (for example, TV repair) ability. S accumulates information from sequence of "checks," seeking correct diagnosis in as few checks as possible (Glaser, Damrin, and Gardner, 1954).

Target Game. Given a number below thirty, blindfolded group of Ss devise strategy for holding up fingers equaling the target number (Zand, 1963).

Teacher's In-Basket Problem. See in-basket problem.

Tourist Problem. Ss think of ways to attract European visitors to U.S.A. (Zagona, Willis, and MacKinnon, 1966).

Tower of Hanoi Problem. See Egyptian pyramid puzzle.

Triangles Problems. Given an orderly triangle-shaped number matrix, S is asked to deduce what number belongs in empty cell (Ray, 1966).

Triangle Maze. S, who may be blindfolded, learns correct sequence of moves in triangle-shaped maze (Erickson, 1962).

Twenty Questions. Same as the old radio program. Given the category *animal, mineral,* or *vegetable,* Ss try to identify the target object in less than twenty questions (Taylor and Faust, 1952).

Two-Armed Bandit. S tries to learn pay-off pattern (Goodnow and Pettigrew, 1956).

Two-String Problem. S devises strategy for tying two strings together. The strings, dangling from ceiling to floor, are spaced so S cannot reach one while holding the other (Maier, 1970). Ray (1966) miniaturized this problem.

Water-Jar Problems. Given three jars with specified volume, S tries to produce a specified amount of water (Bugelski and Huff, 1962; Luchins and Luchins, 1950, 1959).

Word Formation Problem. Given the number of letters in a word, S tries to identify the word by naming as few letters as possible (Battig, 1957).

X-Ray (or *Radiation*) *Problem.* Ss devise strategy for treating internal tumor by radiation without damaging neighboring healthy tissue (Duncker, 1945).

Untitled Test. S tries to find "correct" point on grid in minimum number of choices (Donahoe, 1960).

Untitled Test. S learns to predict which of ten lights will come on next (Banghart, 1959).

Untitled Test. Pattern of eight black or white circles are hidden beneath shutters. S tries to identify total pattern by opening minimum number of shutters (Niemark and Lewis, 1968). Task is said to require diagnostic or logical problem-solving.

Untitled Test. S gives *number* response to *word* stimulus. He must learn from experimenter's "right" or "wrong" that "right" numbers equal number of letters in word minus one (Stein, 1966).

REFERENCES

Advanced Technology Staff, Martin Company. 1961. *Bionics.* A Martin/Orlando Report on Current Missile Engineering Technology.
———. 1963. Bionics I. *Engineering Digest.* Pp. 21–22, 25. (a)
———. 1963. Bionics II. *Engineering Digest.* Pp. 16–18. (b)
Afsar, S., and Koos, E. M. 1970. Behavioral checklist for science students. Kansas City, Mo.: Mid-continent Regional Laboratory.
Alexander, T. 1965. Synectics: Inventing by the madness method. *Fortune* 72 (2): 165–168, 190, 193–194. Reprinted in G. A. Davis and J. A. Scott, eds. 1971. *Training creative thinking.* New York: Holt.
Allen, M. S. 1962. *Morphological creativity.* Englewood Cliffs, N.J.: Prentice-Hall.
———. 1966. *Psycho-dynamic synthesis.* West Nyack, N.Y.: Parker.
Allender, J. S. 1968. *The teaching of inquiry skills to elementary school children.* USOE Cooperative Research Project No. 5-0594, Miami University, Oxford, Ohio.
———. 1969. A study of inquiry activity in elementary school children. *American Educational Research Journal* 6: 543–558.
Allender, D. S., and Allender, J. S. 1965. *I am the mayor.* Diazoed, Miami University, Oxford, Ohio.
Anderson, B., and Johnson, W. 1966. Two methods of presenting information and their effects on problem solving. *Perceptual and Motor Skills* 23: 851–856.

Bahlke, S. J. 1967. A study of the enhancement of creative abilities in elementary school children. Unpublished master's thesis, Purdue University.

———. 1969. Componential evaluation of creativity instructional materials. Unpublished doctoral thesis, Purdue University.

Baker, F. B. 1964. An IPL-V program for concept attainment. *Educational and Psychological Measurement* 24: 119–127.

———. 1965. CASE: A program for simulation of concept learning. *AFIPS Conference Proceedings, 1965 Fall Joint Computer Conference.* Washington, D.C.: Spartan Books.

———. 1967. The internal organization of computer models of cognitive behavior. *Behavioral Science* 12: 156–161.

———. 1968. The development of a computer model of the concept attainment process: A final report. Theoretical Paper No. 16, Wisconsin Research and Development Center for Cognitive Learning, University of Wisconsin.

Banghart, F. 1959. Group structure, anxiety, and problem solving effectiveness. *Journal of Experimental Education* 28: 171–174.

Barron, F. 1958. The psychology of imagination. *Scientific American* 199: 151–166.

———. 1963. *Creativity and psychological health.* Princeton, N.J.: Van Nostrand.

———. 1968. *Creativity and personal freedom.* Princeton, N.J.: Van Nostrand.

———. 1969. *Creative person and creative process.* New York: Holt.

Battig, W. F. 1957. Some factors affecting performance on a word-formation problem. *Journal of Experimental Psychology* 54: 96–104.

Baylor, G. W. 1965. Report on a mating combinations program. SDC Paper No. SP-2150. Systems Development Corporation, Santa Monica, Calif.

Baylor, G. W., and Simon, H. A. 1966. A chess mating combinations program. In *AFIPS Conference Proceedings, 1966 Spring Joint Computer Conference.* Washington, D.C.: Spartan Books.

Berger, R. M.; Guilford, J. P.; and Christensen, P. R. 1957. A factor analytic study of planning. *Psychological Monographs* 71 (6, Whole No. 435).

Blumenfeld, W. 1964. Practice with repetition and reward with-

References

out improvement. *Journal of General Psychology* 71: 313–321.
Bourne, L. E., Jr. 1966. *Human conceptual behavior*. Boston: Allyn & Bacon.
Bourne, L. E., Jr., and Restle, F. 1959. Mathematical theory of concept identification. *Psychological Review* 66: 278–296.
Bower, G., and Trabasso, T. 1963. Reversals prior to solution in concept identification. *Journal of Experimental Psychology* 66: 409–418.
Bruner, J. S. 1957. On going beyond the information given. In *Contemporary approaches to cognition*. Cambridge, Mass.: Harvard University Press. Pp. 41–69.
Bruner, J. S.; Goodnow, J. J.; and Austin, G. A. 1956. *A study of thinking*. New York: Wiley.
Bugelski, B. R., and Huff, E. M. 1962. A note on increasing the efficiency of Luchins' mental sets. *American Journal of Psychology* 75: 665–667.
Burmester, M. A.; Garth, R.E.; Koos, E. M.; and Stothart, J. R. 1970. *Explorations in biology* (Topic 1, Booklet A). Kansas City, Mo.: Mid-continent Regional Educational Laboratory. (a)
———. 1970. *Explorations in biology* (Topic 1, Booklet B). Kansas City, Mo.: Mid-continent Regional Educational Laboratory. (b)
Butts, D. P. 1965. The relationship of problem solving ability and science knowledge. *Science Education* 49: 138–146.
Butts, D. P., and Jones, H. L. 1966. Inquiry training and problem solving in elementary school children. *Journal of Research in Science Teaching* 4: 21–27.
Campbell, D. T. 1960. Blind variation and selective retention in creative thought as in other knowledge processes. *Psychological Review* 67: 380–400.
Carroll, J. B. 1941. A factor analysis of verbal abilities. *Psychometrika* 6: 279–307.
Christensen, P. R., and Guilford, J. P. 1963. An experimental study of verbal fluency factors. *British Journal of Psychology* 16: 1–26.
Clark, C. H. 1958. *Brainstorming*. New York: Doubleday.
Cline, V. B.; Richards, J. M; and Needham, W. E. 1963. Creativity tests and achievement in high school science. *Journal of Applied Psychology* 47: 184–189.
Cofer, C. N. 1957. Reasoning as an associative process: III. The

role of verbal responses in problem solving. *Journal of General Psychology* 57: 55–68.

Comella, T. 1966. Understanding creativity for use in managerial planning. *Automation* 13 (4): 72–76. Reprinted in G. A. Davis and J. A. Scott, eds. 1971. *Training creative thinking.* New York: Holt.

Covington, M. V. 1966. A childhood attitude inventory for problem solving. *Journal of Educational Measurement* 3: 234.

Covington, M. V., and Crutchfield, R. S. 1965. Facilitation of creative problem solving. *Programmed Instruction* 4 (4): 3–5, 10.

Covington, M. V.; Crutchfield, R. S.; and Davies, L. B. 1966. *The productive thinking program. Series one: General problem solving.* Berkeley: Brazelton Printing Company. Augmented edition by Covington, M. V.; Crutchfield, R. S.; Davies, L. B.; and Olton, R. M. 1972. Columbus, Ohio: Charles E. Merrill Publishing Company.

Cowen, E. L.; Wiener, M.; and Hess, J. 1953. Generalization of problem solving rigidity. *Journal of Consulting Psychology* 17: 100–103.

Crawford, R. P. 1937. *Think for yourself.* Burlington, Vt.: Fraser.

———. 1948. *How to get ideas.* Lincoln, Neb.: University Associates.

———. 1954. *Techniques of creative thinking.* New York: Hawthorn.

———. 1964. *Direct creativity.* Burlington, Vt.: Fraser.

Crutchfield, R. S., and Covington, M. V. 1965. Programmed instruction and creativity. *Programed Instruction* 4 (4): 1–2, 8–10.

Davis, G. A. 1965. Stimulus and response variables in human problem solving. Unpublished doctoral dissertation, University of Wisconsin.

———. 1966. Current status of research and theory in human problem solving. *Psychological Bulletin* 66: 36–54.

———. 1967. Detrimental effects of distraction, additional response alternatives, and longer response chains in solving switch-light problems. *Journal of Experimental Psychology* 73: 45–55.

———. 1969. Training creativity in adolescence: A discussion of strategy. In R. E. Grinder, ed., *Studies in adolescence*

References

II. New York: Macmillan. Pp. 538–545. Reprinted in G. A. Davis and J. A. Scott, eds. 1971. *Training creative thinking.* New York: Holt.

———. 1970. Problems in assessing the effectiveness of creative thinking. Presented in a symposium entitled, "Assessing creativity: Progress in both directions," American Educational Research Association, Minneapolis, March. (a)

———. 1970. Training creative thinking: In the suburbs and the inner city. Paper presented at the Eighth Creativity Conference, Buffalo, June. (b)

Davis, G. A., and Houtman, S. E. 1968. *Thinking creatively: A guide to training imagination.* Wisconsin Research and Development Center for Cognitive Learning, University of Wisconsin.

Davis, G. A.; Houtman, S. E.; Warren, T. F.; and Roweton, W. E. 1969. A program for training creative thinking: I. Preliminary field test. Technical Report No. 104, Wisconsin Research and Development Center for Cognitive Learning, University of Wisconsin.

Davis, G. A.; Houtman, S. E.; Warren, T. F.; Roweton, W. E.; Mari, S. K.; and Belcher, T. L. 1972. A program for training creative thinking: II. Inner city evaluation. Technical Report no. 224, Wisconsin Research and Development Center for Cognitive Learning, University of Wisconsin.

Davis, G. A., and Manske, M. E. 1968. Effects of prior serial learning of solution words upon anagram problem solving: II. A serial position effect. *Journal of Experimental Psychology* 77: 101–104.

Davis, G. A., and Roweton, W. E. 1968. Using idea checklists with college students: Overcoming resistance. *Journal of Psychology* 70: 221–226.

Davis, G. A.; Roweton, W. E.; Train, A. J.; Warren, T. F.; and Houtman, S. E. 1969. Laboratory studies of creative thinking techniques: The checklist and morphological synthesis methods. Technical Report No. 94, Wisconsin Research and Development Center for Cognitive Learning, University of Wisconsin.

Davis, G. A., and Scott, J. A., eds. 1971. *Training Creative Thinking.* New York: Holt.

DeGroot, A. D. 1965. *Thought and choice in chess.* The Hague: Mouton.

DeGroot, A. D. 1966. Perception and memory versus thought: Some old ideas and recent findings. In B. Kleinmuntz, ed., *Problem solving: Research, method and theory.* New York: Wiley. Pp. 19–50.

Dewey, J. 1933. *How we think.* New York: Heath.

———. 1938. *Logic: The theory of inquiry.* New York: Holt.

DiPego, G. 1970. *Saturday subway ride.* Wisconsin Research and Development Center for Cognitive Learning, University of Wisconsin.

Domino, G. 1970. Identification of potentially creative persons from the Adjective Check List. *Journal of Consulting and Clinical Psychology* 35: 48–51.

Donahoe, J. W. 1960. The effect of variations in the form of feedback on the efficiency of problem solving. *Journal of Experimental Psychology* 60: 193–199.

Duncker, K. 1945. On problem solving. *Psychological Monographs* 58 (5, Whole No. 270). Reprinted in part in W. Kessen and G. Mandler, eds. 1964. *Thinking: From association to Gestalt.* New York: Wiley. Pp. 262–298.

Dunnette, M. D.; Campbell, J.; and Jaastad, K. 1963. Effect of group participation on brainstorming effectiveness for two industrial samples. *Journal of Applied Psychology* 47: 30–37.

Edwards, M. W. 1968. A survey of problem-solving courses. *Journal of Creative Behavior* 2: 33–51.

Elms, A. C. 1966. Influence of fantasy ability on attitude change through role playing. *Journal of Personality and Social Psychology* 4: 36–43.

Erickson, S. C. 1962. Studies in the abstraction process. *Psychological Monographs* 76 (18, Whole No. 537).

Evans, T. 1968. A program for solution of geometry analogy intelligence test items. In M. Minsky, ed., *Semantic information processing.* Cambridge, Mass.: Massachusetts Institute of Technology Press.

Fabun, D. 1968. You and creativity. *Kaiser Aluminum News* 25 (3).

Feigenbaum, E. A., and Feldman, J., eds. 1963. *Computers and thought.* New York: McGraw-Hill.

Feldhusen, J. F.; Bahlke, S. J.; and Treffinger, D. J. 1969. Teaching creative thinking. *Elementary School Journal* 70: 48–53.

Feldhusen, J. F.; Treffinger, D. J.; and Bahlke, S. J. 1970. Developing creative thinking: The Purdue creativity program. *Journal of Creative Behavior* 4: 85–90.

Flescher, I. 1963. Anxiety and achievement of intellectually gifted and creatively gifted children. *Journal of Psychology* 56: 251–268.

Frandsen, A. N., and Holder, J. R. 1969. Spatial visualization in complex verbal problems. *Journal of Psychology* 73: 229–233.

Frank, F. 1966. Perception and language in conservation. In J. S. Bruner, ed., *Studies in cognitive growth*. New York: Wiley.

Frederiksen, N.; Saunders, D. R.; and Wand, B. 1957. The in-basket test. *Psychological Monographs* 70 (9, Whole No. 438).

French, J. W. 1951. The description of aptitude and achievement tests in terms of rotated factors. *Psychometric Monographs*, No. 5.

Gentile, J. R.; Kessler, D. K.; and Gentile, P. K. 1969. Process of solving analogy items. *Journal of Educational Psychology* 60: 494–502.

Gerard, R. W. 1952. The biological basis of imagination. In B. Ghiselin, ed., *The creative process*. Berkeley: University of California Press.

Gerlernter, H. 1959. Realization of a geometry theorem-proving machine. *Proceedings of the International Conference on Information Processing*. Paris: UNESCO House. Pp. 273–282.

Getzels, J. W., and Jackson, P. W. 1962. *Creativity and intelligence*. New York: Wiley.

Glaser, R.; Damrin, D. E.; and Gardner, F. M. 1954. The Tab Item: A technique for measurement of proficiency in diagnostic problem solving tasks. *Educational and Psychological Measurement* 14: 283–293.

Glucksberg, S. 1964. Problem solving: Response competition and the influence of drive. *Psychological Reports* 15: 939–942.

Goldiamond, I. 1966. Perception, language, and conceptualization rules. In B. Kleinmuntz, ed., *Problem solving: Research, method and theory*. New York: Wiley. Pp. 183–224.

Gollin, E. S., and Shirk, E. J. 1966. A developmental study of oddity-problem learning in young children. *Child Development* 37: 213–217.

Goodnow, J., and Pettigrew, T. 1956. Some sources of difficulty in solving simple problems. *Journal of Experimental Psychology* 51: 385–392.

Gordon, W. J. J. 1961. *Synectics*. New York: Harper & Row.

———. 1969. *The metaphorical way of learning and knowing.* Prepublication edition. Cambridge, Mass.: Synectics.

Gregg, L. W., and Simon, H. A. 1967. An information-processing explanation of one-trial and incremental learning. *Journal of Verbal Learning and Verbal Behavior* 6: 780–787.

Green, C. C. 1969. Theorems proving by resolution as a basis for question answering systems. In B. Meltzer and D. Michie, eds., *Machine intelligence 4*. Edinburgh: Edinburgh University Press. (a)

———. 1969. Application of theorem proving to problem solving. *Proceedings of the International Joint Conference on Artificial Intelligence*. Bedford, Mass.: MITRE Corp. Pp. 211–240. (b)

Guilford, J. P. 1962. Creativity: Its measurement and development. In S. J. Parnes and H. F. Harding, eds., *A source book for creative thinking*. New York: Scribner's. Pp. 151–168.

———. 1967. *The nature of human intelligence*. New York: McGraw-Hill.

Guilford, J. P., and Merrifield, P. R. 1960. The structure of intellect model: Its uses and implications. Reports from the Psychological Laboratory, No. 24, The University of Southern California.

Guilford, J. P.; Wilson, R. C.; and Christensen, P. R. 1952. A factor analytic study of creative thinking: II. Administration of tests and analysis of results. Reports from the Psychological Laboratory, No. 8, The University of Southern California.

Guilford, J. P.; Wilson, R. C.; Christensen, P. R.; and Lewis, D. J. 1951. A factor-analytic study of creative thinking: I. Hypotheses and descriptions of tests. Reports from the Psychological Laboratory, No. 4, The University of Southern California.

Hadamard, J. 1945. *An essay on the psychology of invention in the mathematical field*. New York: Dover.

Hayes, J. R. 1966. Memory, goals, and problem solving. In B. Kleinmuntz, ed., *Problem solving: Research, method and theory*. New York: Wiley. Pp. 149–170.

Haygood, R. C., and Bourne, L. E., Jr. 1965. Attribute- and rule-learning aspects of conceptual behavior. *Psychological Review* 72: 175–195.

References

Henle, M. 1962. On the relation between logic and thinking. *Psychological Review* 69: 366–378.

Hoffman, L. R. 1961. Conditions for creative problem solving. *Journal of Psychology* 52: 429–444.

Holland, J. L., and Baird, L. C. 1968. The Preconscious Activity Scale: The development and validation of an originality measure. *Journal of Creative Behavior* 2: 214–223.

Houtman, S. E. 1970. *Write? Right!* Wisconsin Research and Development Center for Cognitive Learning, University of Wisconsin. (a)

———. 1970. *Write? Right! Teachers' Manual*. Wisconsin Research and Development Center for Cognitive Learning, University of Wisconsin. (b)

Hull, C. L. 1934. The concept of the habit-family-hierarchy and maze learning: Part I. *Psychological Review* 41: 33–54.

———. 1952. *A behavior system*. New Haven: Yale University Press.

Hunt, E. B. 1962. *Concept learning: An information processing problem*. New York: Wiley.

———. 1968. Computer simulation: Artificial intelligence studies and their relevance to psychology. *Annual Review of Psychology* 19: 135–168.

———. 1970. What kind of computer is man? Technical Report No. 70-1-01, Department of Psychology, University of Washington.

Hunt, E. B., and Hovland, C. I. 1961. Programming a model of human concept formation. *Proceedings of the Western Joint Computer Conference* 19: 145–155.

Hyman, R. 1964. Creativity and the prepared mind: The role of information and induced attitudes. In C. W. Taylor, ed., *Widening Horizons in Creativity*. New York: Wiley. Pp. 69–79.

Johnson, D. M. 1955. *The psychology of thought and judgment*. New York: Harper.

———. 1960. Serial analysis of thinking. *Annals of the New York Academy of Sciences* 91: 66–75.

———. 1962. Serial Analysis of verbal analogy problems. *Journal of Educational Psychology* 53: 86–88.

———. 1966. Solution of anagrams. *Psychological Bulletin* 66: 371–384.

Johnson, D. M., and Zerbolio, D. J. 1964. Relations between production and judgment of plot titles. *American Journal of Psychology* 77: 99–105.

Johnson, E. S. 1964. An information processing model of one kind of problem solving. *Psychological Monographs* 78 (4, Whole No. 581).

Judson, A., and Cofer, C. N. 1956. Reasoning as an associative process: I. Direction in a simple verbal problem. *Psychological Reports* 2: 469–476.

Kaplan, I. T., and Carvellas, T. 1969. Searching for words in letter sets of varying size. *Journal of Experimental Psychology* 82: 377–380.

Katona, G. 1940. *Organizing and memorizing.* New York: Columbia University Press.

Kendler, H. H. 1961. Problems in problem solving research. In *Current trends in psychological theory: A bicentennial program.* Pittsburgh: University of Pittsburgh Press.

Kendler, H. H., and Kendler, T. S. 1962. Vertical and horizontal processes in problem solving. *Psychological Review* 69: 1–16.

Kendler, T. S. 1961. Concept formation. *Annual Review of Psychology* 12: 447–472.

Kendler, T. S., and Kendler, H. H. 1961. Inferential behavior in children: II. The influence of order of presentation. *Journal of Experimental Psychology* 61: 442–448.

Kilpatrick, J. 1969. Problem solving in mathematics. *American Educational Research Journal* 39: 523–534.

Kingsley, H. L., and Garry, R. 1957. *The nature and conditions of learning.* 2d ed. Englewood Cliffs, N.J.: Prentice-Hall.

Köhler, W. 1925. *The mentality of apes.* New York: Harcourt Brace.

Koos, E. M. 1969. A report on developmental studies of the first in a series of measures of inquiry skill in biology, "Explorations in biology, Topic 1." Technical Report No. 1, Midcontinent Regional Educational Laboratory, Kansas City, Mo.

Linker, E., and Ross, B. M. 1963. Memory and hypotheses in solving alternation problems with random competition. *Psychological Reports* 12: 783–797.

Lorge, I.; Tuckman, J.; Aikman, L.; Spiegel, J.; and Moss, G. 1956. The adequacy of written reports in problem solving by teams and individuals. *Journal of Social Psychology* 43: 65–74.

Luchins, A. S. 1942. Mechanization in problem solving—the

References

effect of Einstellung. *Psychological Monographs* 54 (6, Whole No. 248).

Luchins, A. S., and Luchins, E. H. 1950. New experimental attempts at preventing mechanization in problem solving. *Journal of General Psychology* 42: 279–297.

———. 1959. *Rigidity of behavior*. Eugene, Ore.: University of Oregon Books.

Lyda, L., and Fillanbaum, S. 1964. Dogmatism and problem solving: An examination of the Denny Doodlebug problem. *Psychological Reports* 14: 99–102.

Maier, N. R. F. 1930. Reasoning in humans: I. On direction. *Journal of Comparative and Physiological Psychology* 10: 114–143.

———. 1933. An aspect of human reasoning. *British Journal of Psychology* 24: 144–155.

———. 1945. Reasoning in humans: III. The mechanisms of equivalent stimuli and of reasoning. *Journal of Experimental Psychology* 35: 349–360.

———. 1970. *Problem solving and creativity: In individuals and groups*. Belmont, Cal.: Brooks/Cole.

Maltzman, I. 1955. Thinking: From a behavioristic point of view. *Psychological Review* 62: 275–286.

———. 1960. On the training of originality. *Psychological Review* 67: 229–242.

Manske, M. E., and Davis, G. A. 1968. Effects of simple instructional biases upon performance in the unusual uses test. *Journal of General Psychology* 79: 25–33.

Marks, M. R. 1951. Problem solving as a function of the situation. *Journal of Experimental Psychology* 41: 74–80.

Mason, J. G. 1960. *How to be a more creative executive*. New York: McGraw-Hill.

Mayzner, M. S., and Tresselt, M. E. 1958. Anagram solution times: A function of letter order and word frequency. *Journal of Psychology* 56: 376–379.

Merrifield, P. R. 1971. *Thinking interest inventory*. Unpublished test, Department of Educational Psychology, New York University.

Miller, G. A. 1956. The magical number seven, plus or minus two: Some limits on our capacity for processing information. *Psychological Review* 63: 81–97.

Moore, O. K., and Anderson, S. B. 1954. Modern logic and

tasks for experiments on problem solving behavior. *Journal of Psychology* 38: 151–160.

Myers, R. E., and Torrance, E. P. 1964. *Invitations to thinking and doing*. Boston: Ginn.

———. 1965. *Can you imagine?* Boston: Ginn. (a)

———. 1965. *Invitations to speaking and writing creatively*. Boston: Ginn. (b)

———. 1966. *For those who wonder*. Boston: Ginn. (a)

———. 1966. *Plots, puzzles and ploys*. Boston: Ginn. (b)

———. 1968. *Stretch*. Minneapolis: Perceptive.

Newell, A.; Shaw, J. C.; and Simon, H. A. 1957. Empirical explorations with the logic theory machine. *Proceedings of the Western Joint Computer Conference* 11: 218–230.

———. 1958. Elements of a theory of human problem solving. *Psychological Review* 65: 151–166. Reprinted in R. C. Anderson and D. P. Ausubel, eds. 1965. *Readings in the psychology of cognition*. New York: Holt. Pp. 133–158.

———. 1962. The process of creative thinking. In H. E. Gruber, G. Terrell, and M. Wertheimer, eds., *Contemporary approaches to creative thinking*. New York: Atherton. Pp. 63–119.

———. 1963. Chess playing and the problem of complexity. In E. A. Feigenbaum and J. Feldman, eds., *Computers and thought*. New York: Wiley. Pp. 39–70.

Newell, A., and Simon, H. A. 1961. Computer simulation of human thinking. *Science* 134: 2011–2017.

———. 1965. An example of human chess play in the light of chess playing programs. In N. Wiener and P. Schade, eds., *Progress in biocybernetics*. Amsterdam, N.Y.: Elsevier.

Niemark, E. D., and Lewis, N. 1968. Development of logical problem solving: A one-year retest. *Child Development* 39: 527–536.

Odom, R. D. 1967. Problem solving strategies as a function of age and socioeconomic level. *Child Development* 38: 747–752.

Olton, R. M., and Crutchfield, R. S. 1969. Developing the skills of productive thinking. In P. Mussen, J. Langer, and M. V. Covington, eds., *New directions in developmental psychology*. New York: Holt. Pp. 68–91. Reprinted in G. A. Davis and J. A. Scott, eds. 1971. *Training creative thinking*. New York: Holt.

Olton, R. M.; Wardrop, J.; Covington, M. V.; Goodwin, W.;

References

Crutchfield, R. S.; Klausmeier, H. J.; and Ronda, T. 1967. The development of productive thinking skills in fifth-grade children. Technical Report No. 34, Wisconsin Research and Development Center for Cognitive Learning, University of Wisconsin.

Osborn, A. F. 1942. *How to "think up."* New York: McGraw-Hill.

———. 1948. *Your creative power.* New York: Scribner's.

———. 1952. *Wake up your mind.* New York: Scribner's.

———. 1963. *Applied Imagination.* 3d ed. New York: Scribner's.

Paige, J. M., and Simon, H. A. 1966. Cognitive processes in solving algebra word problems. In B. Kleinmuntz, ed., *Problem solving: Research, method and theory.* New York: Wiley. Pp. 51–119.

Paivio, A. 1969. Mental imagery in associative learning. *Psychological Review* 76: 241–263.

Papanek, V. J. 1969. Tree of life: Bionics. *Journal of Creative Behavior* 3: 5–15.

Parnes, S. J. 1961. Effects of extended effort in creative problem solving. *Journal of Educational Psychology* 53: 117–122.

———. 1967. *Creative behavior guidebook.* New York: Scribner's.

Parnes, S. J., and Harding, H. F., eds. 1962. *A source book for creative thinking.* New York: Scribner's.

Polya, G. *How to solve it.* 1957. 2d ed. Garden City, N.Y.: Doubleday Anchor.

Prince, G. 1968. The operational mechanism of synectics. *Journal of Creative Behavior* 2: 1–13.

Pryor, K. W.; Haag, R.; and O'Reilly, J. 1969. The creative porpoise: Training for novel behavior. *Journal of the Experimental Analysis of Behavior* 12: 653–661.

Raaheim, K. 1963. The concept of the deviant situation. *Psychological Reports* 13: 174.

Ray, W. S. 1966. Originality in problem solving as affected by single-*versus* multiple-solution training problems. *Journal of Psychology* 64: 107–112.

Reitman, W. 1965. *Cognition and thought.* New York: Wiley.

Reitman, W.; Grove, R. B.; and Shoup, R. G. 1964. Argus: An information processing model of thinking. *Behavioral Science* 9: 270–280.

Restle, F., and Davis, J. H. 1962. Success and speed of problem solving by individuals and groups. *Psychological Review* 69: 520–536.

Robinson, J. A. 1965. A machine oriented logic based on the resolution principle. *Journal of the Association for Computing Machinery* 12: 23–41.

———. 1969. Mechanizing higher order logic. In B. Meltzer and E. Michie, eds., *Machine intelligence 4*. Edinburgh: Edinburgh University Press.

Robinson, W. L. T. 1969. Taped-creativity-series versus conventional teaching and learning. Unpublished master's thesis, Atlanta University.

Rohwer, W. D., Jr., and Ammon, M. S. 1971. Elaboration training and paired-associate learning efficiency in children. *Journal of Educational Psychology* 62: 376–383.

Ronning, R. R. 1965. Anagram solution times: A function of the "ruleout" factor. *Journal of Experimental Psychology* 69: 35–39.

Rookey, T. J. 1967. Pennsylvania Assessment of Creative Tendency (PACT). Unpublished test, Department of Public Instruction, Commonwealth of Pennsylvania, Harrisburg, Pa.

Saugstad, P., and Raaheim, K. 1960. Problem solving, past experience and availability of functions. *British Journal of Psychology* 51: 97–104.

Schulz, R. W. 1960. Problem solving behavior and transfer. *Harvard Educational Review* 30: 61–77. Reprinted in R. F. Grose and R. C. Birney, eds. 1963. *Transfer of learning*. Princeton, N.J.: Van Nostrand. Pp. 160–180.

Shaw, M. E. 1958. Some effects of irrelevant information upon problem solving by small groups. *Journal of Social Psychology* 47: 33–37.

Sheerer, M. 1963. Problem solving. *Scientific American* 208 (4): 118–128.

Shulman, L. S. 1965. Seeking styles and individual differences in patterns of inquiry. *School Review* 73: 258–266.

———. 1967. The study of individual inquiry behavior. Presented as part of the symposium *Studies of the inquiry process: Problems of theory* at the meeting of the American Psychological Association, Washington, D.C.

Shulman, L. S.; Loupe, M. J.; and Piper, R. M. 1968. *Studies of the inquiry process: Inquiry patterns of students in teacher-training programs*. East Lansing: Educational Publications Services, Michigan State University.

References

Simberg, A. L. 1964. *Creativity at work*. Boston: Industrial Education Institute.

Simon, H. A., and Barenfeld, M. 1969. Information-processing analysis of perceptual processes in problem solving. *Psychological Review* 76: 473–483.

Simon, H., and Kotovsky, K. 1963. Human acquisition of concepts for sequential patterns. *Psychological Review* 70: 534–546.

Simon, H. A., and Newell, A. 1956. Models, their uses and limitations. In L. D. White, ed., *The state of the social sciences*. Chicago: University of Chicago Press. Pp. 66–83.

———. 1961. Computer simulation of human thinking and problem solving. *Datamation* 7: 18–20.

Simon, H., and Simon, P. 1962. Trial and error search in solving difficult problems: Evidence from the game of chess. *Behavioral Science* 7: 425–429.

Skinner, B. F., 1966. An operant analysis of problem solving. In B. Kleinmuntz, ed., *Problem solving: Research, method and theory*. New York: Wiley. Pp. 225–257.

Slagle, J. R. 1963. A heuristic program that solves symbolic integration problems in freshman calculus. *Journal of the Association for Computing Machinery* 10: 507–520.

Slagle, J. R., and Dixon, J. 1969. Experiments with some programs that search game trees. *Journal of the Association for Computing Machinery* 16: 189–207.

Staats, A. W. 1966. An integrated-functional learning approach to complex human behavior. In B. Kleinmuntz, ed., *Problem solving: Research, method and theory*. New York: Wiley. Pp. 259–339.

———. 1968. *Learning, language, and cognition*. New York: Holt.

Staats, A. W., and Staats, C. K. 1963. *Complex human behavior*. New York: Holt.

Stein, L. S. 1966. Conscious mediating processes in a problem-solving task. *Journal of Experimental Psychology* 71: 212–217.

Stevenson, H. W., and Odom, R. D. 1964. Children's behavior in a probabilistic situation. *Journal of Experimental Psychology* 68: 260–268.

Stevenson, H. W., and Weir, M. W. 1963. The role of age and verbalization in probability learning. *American Journal of Psychology* 76: 299–305.

Suchman, J. R. 1960. Inquiry training in the elementary school. *Science Teacher* 27: 42–47.

———. 1963. Developing inquiry skills in children. *The Instructor* 72 (7).

———. 1966. A model for the analysis of inquiry. In H. J. Klausmeier and C. W. Harris, eds., *Analyses of concept learning*. New York: Academic Press. Pp. 177–187. (a)

———. 1966. *Inquiry development program in physical science*. Chicago: Science Research Associates. (b)

———. 1967. *Inquiry box: An educational device for developing inquiry*. Chicago: Science Research Associates.

———. 1968. *Inquiry development program in earth science*. Chicago: Science Research Associates.

———. 1969. *Evaluating inquiry in physical science*. Chicago: Science Research Associates.

Suedfeld, P.; Glucksberg, S.; and Vernon, J. 1967. Sensory deprivation as a drive operation: Effects upon problem solving. *Journal of Experimental Psychology* 75: 166–169.

Taft, R. 1969. Peak experiences and ego permissiveness: An exploratory factor study of their dimensions. *Acta Psychologica* 29: 35–64.

Taft, R., and Gilchrist, M. B. 1970. Creative attitudes and creative productivity: A comparison of two aspects of creativity among students. *Journal of Educational Psychology* 61: 136–143.

Taylor, C. W. 1947. A factorial study of fluency in writing. *Psychometrika* 12: 239–262.

Taylor, C. W., and Barron, F., eds. 1963. *Scientific creativity: Its recognition and development*. New York: Wiley.

Taylor, D. W.; Berry, P. C.; and Block, C. N. 1958. Does group participation when using brainstorming facilitate or inhibit creative thinking? *Administrative Science Quarterly* 3: 23–47.

Taylor, D. W., and Faust, W. L. 1952. Twenty questions: Efficiency in problem solving as a function of size of group. *Journal of Experimental Psychology* 44: 360–368.

Thorndike, E. L. 1898. Animal intelligence: An experimental study of the associative processes in animals. *Psychological Review Monograph Supplements* 2 (6, Whole No. 8).

———. 1911. *Animal intelligence*. New York: Macmillan.

Thurstone, L. L. 1944. A factorial study of perception. *Psychometrika Monographs*, No. 4.

References

Tichomirov, O. K., and Poznyanskaya, E. D. 1966–1967. An investigation of visual search as a means of analyzing heuristics. *Soviet Psychology* 5 (Winter): 2–15. Trans. from *Voprosy Psikhologii.* 1966. 2 (4): 39–53.

Torrance, E. P. 1962. *Guiding creative talent.* Englewood Cliffs, N. J.: Prentice-Hall.

———. 1965. *Rewarding creative behavior.* Englewood Cliffs, N.J.: Prentice-Hall.

———. 1971. Some validity studies of two brief screening devices for studying the creative personality. *Journal of Creative Behavior*, 5: 94–103.

Torrance, E. P., and Myers, R. E. 1970. *Creative learning and teaching.* New York: Dodd.

Train, A. J. 1967. Attribute listing and use of a checklist: A comparison of two techniques for stimulating creative thinking. Unpublished master's thesis, University of Wisconsin.

Treffinger, D. J., and Ripple, R. E. 1969. Developing creative problem solving abilities and related attitudes through programmed instruction. *Journal of Creative Behavior* 3: 105–110.

Value Engineering Department of Lockheed-Georgia. 1971. Value engineering. In G. A. Davis and J. A. Scott, eds., *Training Creative Thinking.* New York: Holt.

Wallach, M. A., and Kogan, N. 1965. *Modes of thinking in young children.* New York: Holt.

Wallas, G. 1926. *The art of thought.* New York: Harcourt, Brace & World.

Warren, T. F. 1970. Creative thinking techniques: four methods of stimulating original ideas in sixth grade children. Unpublished doctoral dissertation, University of Wisconsin.

Warren, T. F., and Davis, G. A. 1969. Techniques for creative thinking: An empirical comparison of three methods. *Psychological Reports* 25: 207–214.

Wason, P. C. 1968. On the failure to eliminate hypotheses—a second look. In P. C. Wason and P. N. Johnson-Laird, eds., *Thinking and reasoning.* Baltimore: Penguin Books. Pp. 165–174.

WBAA. *Creative thinking: The American Pioneers.* 1966. (Audio tapes and a manual for teachers.) West Lafayette, Ind.: Purdue University School of the Air.

Weisskopf-Joelson, E., and Eliseo, T. S. 1961. An experimental

study of the effectiveness of brainstorming. *Journal of Applied Psychology* 45: 45–49.
Welch, L. 1946. Recombination of ideas in creative thinking. *Journal of Applied Psychology* 30: 638–643.
Wertheimer, M. 1945. *Productive thinking*. New York: Harper.
Whitehead, A. N., and Russell, B. 1925. *Principia mathematica.* Vol. I. 2d ed. Cambridge: Cambridge University Press.
Witkin, H. A.; Dyk, R. B.; Faterson, H. F.; Goodenough, D. R.; and Karp, S. A. 1962. *Psychological differentiation*. New York: Wiley.
Wittrock, M. C. 1963. Verbal stimuli in concept formation: Learning by discovery. *Journal of Educational Psychology* 54: 183–190.
Woodworth, R. S., and Schlosberg, H. 1954. *Experimental psychology*. New York: Holt.
Zagona, S. V.; Willis, J. E.; and MacKinnon, W. J. 1966. Group effectiveness in creative problem solving tasks: An examination of relevant variables. *Journal of Psychology* 62: 111–137.
Zand, D. E., and Costello, T. 1963. Effect of problem variation on group problem solving efficiency under constrained communication. *Psychological Reports* 13: 219–224.
Zwicky, F. 1957. *Morphological astronomy*. Berlin: Springer-Verlag.

INDEX

abilities, 8, 9, 157, 164
Advanced Technology Staff, Martin Company, 128, 130–131, 181
Afsar, S., 85, 181
Aikman, L., 190
Alexander, T., viii, 121, 132, 181
algorithmic computer programs, 61–62, 65, 67, 70, 162–163
Allender, D. S., 83–84, 181
Allender, J. S., 80, 83–84, 181
Allen, M. S., 105–106, 181
Ammon, M. S., 74, 194
anagram problems, 5, 9, 12, 15, 17, 20, 22, 23, 47, 49, 50, 51–53, 79, 174, 175, 177
analogy programs, 70
analogy strategies, 122–125, 127, 175
Anderson, S. B., 178, 179, 181, 191
Applied Imagination (Osborn), 7, 103
Argus program, 70
Aristotle, 58
Arnold, John, 109
Artificial Geometer, 70
artificial intelligence, 61, 162
attitudes, 8–9, 102, 134–135, 136, 157
attribute listing, 6, 8, 9, 76, 103, 104–105, 106, 136, 139, 163
Austin, G. A., 37, 45, 183

Bahlke, S. J., 150, 153, 154, 155, 182, 186
Bain, Alexander, 42
Baird, L. C., 173, 189
Baker, F. B., 70, 182

ball-transfer problem, 47, 175, 177
Banghart, F., 182
Barenfeld, M., 67, 68, 69, 195
Barron, F., viii, 25, 91, 168, 169, 170, 171, 182, 196
Battig, W. F., 179, 182
Baylor, G. W., 68, 71, 182
Behavioral Checklist for Science Students (BCL), 85
Belcher, T. L., 140, 149, 185
Berger, R. M., 168, 182
Berry, P. C., 101, 196
bionics, 8, 76, 120, 128–131, 136, 163
Block, C. N., 101, 196
Blumenfeld, W., 175, 182
Bourne, L. E., Jr., vii, viii, 45, 48, 174, 183, 188
Bower, G., 45, 183
brainstorming, viii, 6–7, 8, 9, 17, 76, 90–92, 95–96, 101–102, 126, 128, 135, 163; deferred judgment, 90, 92, 95, 101; defining problems, 97–98; evaluating ideas, 98–100; fluency, 99, 100, 151; ground rules, 92–95; *Phillips 66* technique, 96; stop-and-go, 96
Braun, Werner von, 17
Bruner, Jerome S., 6, 28, 29, 37–39, 45, 47, 162, 163, 183
Buck, Pearl, 17
Bugelski, B. R., 179, 183
Burmeister, M. A., 85, 183
Butts, D. P., 80, 84, 163, 183

Campbell, Donald T., 42, 47–48, 183

199

Campbell, J., 101, 186
candle problems, 20, 47, 175
Can You Imagine? (Idea Book), 151
Carroll, J. B., 169, 183
Carvellas, T., 177, 190
cats, and puzzle box, 41, 42, 48, 49, 53–54, 59, 178
chaining, in computer simulation, 64, 65
chain responses, 6, 55
chains, of associations, 40, 56–57
checkerboard method, 139, 140
checklists, *see* idea checklists
chess, computer playing, 61, 62, 67–69, 71; problems, 79
chimpanzee studies, 30–33, 43, 162
Christensen, P. R., 168, 171, 172, 182, 183, 188
Clark, C. H., 19, 92, 95, 183
Cline, V. B., 170, 183
coded equivalence-class, 38–39
coding, higher-order, 6, 28, 47, 73, 162, 163; models and theories, 38–39
Cofer, C. N., 51, 56, 176, 183, 190
cognitive restructuring, 27, 30, 162
cognitive theories, in psychology, 27–28, 29, 39, 72–73
Comella, T., 184
computer-simulation problem solving, 6, 130–131, 162; algorithmic vs. heuristic programs, 61–62, 65, 67, 70, 162–163; analogy programs, 70; Argus program, 70; Artificial Geometer, 70; chaining, 64, 65; chess playing, 61, 62, 67–69, 71; concept learning, 70; creativity, 71–72; General Problem Solver, 66–67, 69; Logic Theorist (LT), 64–65, 66, 67, 69, 71–72; model, 60, 80; Perceiver, 67–69; programming, 62–64; Saint program, 70
concept learning, 37–39, 45, 70, 174

concept problems, 20, 22, 49
conformity, in problem solving, 33, 161
constructive discontent, 135
Cornell Aeronautical Laboratory, 131
Costello, T., 198
Covington, M. V., 150, 153, 155, 156, 164, 172, 184, 192
Cowen, E. L., 175, 184
Crawford, Robert P., viii, 6, 103–105, 119, 184
Creative Education Foundation, viii, 6
Creative Thinking: The American Pioneers (Purdue), 153, 164
creativity, 8–9, 10, 14, 24, 55–56, 71–72, 119, 136, 148, 156–157, 163
Creativity and Personal Freedom (Barron), 91
Creativity and Psychological Health (Barron), 91
Crutchfield, R. S., 150, 153, 155, 156, 172, 184, 192, 193
Cunnington, B. F., 157, 166
Cybertron, 130

Damrin, D. E., 179, 187
Davies, L. B., 150, 155, 184
Davis, Gary A., 11, 22, 23, 43, 51–52, 94, 100, 102, 106, 108, 111, 112, 113, 114, 119, 125–126, 132, 134, 135, 136, 138, 139, 140, 149, 151, 165, 169, 173, 176, 178, 184, 185, 191, 197
Davis, J. H., 47, 193
deferred judgment, in brainstorming, 90, 92, 95, 101
DeGroot, A. D., 69, 175, 185
detour problem, 30, 31, 32, 49, 56, 176
Dewey, J., 13, 16, 79, 80, 163, 186
DiPego, G., 108, 141, 186
direct analogy, in synectics, 122–125, 127, 139
direction, in problem solving, 35, 73, 162, 163

Index

divergence, vs. convergence, in problem solving, 22, 25, 50, 53
Dixon, J., 70, 195
Domino, G., 186
Donahoe, J. W., 46, 186
dot and lines problem, 35–36
Duncker, K., 33, 47, 48, 156, 179, 186
Dunnette, M. D., 101, 186
Dyk, R. B., 169, 198

Early Explorers (Purdue), 153–154
Edwards, M. W., 11, 77, 186
Eliseo, T. S., 101, 197
Ellington, Duke, 17
Elms, A. C., 172, 186
Erickson, S. C., 46–47, 179, 186
Evans, T., 71, 186
Explorations in Biology (EIB), 85–86

Fabun, D., 21, 25, 186
Faterson, H. F., 169, 198
Faust, W. L., 179, 196
Feigenbaum, E. A., vii, 28, 186
Feldhusen, J. F., 150, 153, 154, 155, 156, 157, 164, 186
Feldman, J., vii, 186
Fighters for Freedom (Purdue), 154
Fillanbaum, S., 176, 191
fixation, 5, 18, 19, 24, 25, 35–36, 48, 57, 73
Flescher, I., 172, 187
fluency, 99, 100, 151
force fit, 127
For Those Who Wonder (Idea Book), 152
Fox, H. H., 21
Frandsen, A. N., 179, 187
Frank, F., 175, 187
Frederickson, N., 89, 186
French, J. W., 168, 169, 171, 187
Frontiers in Health and Science (Purdue), 154
Frontiers in Transportation and Communication (Purdue), 154

Gardner, F. M., 179, 187
Garry, R., 16, 190
Garth, R. E., 85, 183
General Electric, viii, 6
General Problem Solver, 66–67, 69
Gentile, J. R., and P. K., 175, 187
Gerard, R. W., 20–21, 187
Gerlernter, H., 70, 187
Gestalt psychology, 5, 6, 18–19, 27–28, 29, 39, 40, 73, 80, 88, 162
Getzels, J. W., 168, 169, 172, 187
Gilchrist, M. B., 167, 173, 196
Glaser, R., 179, 187
Glucksberg, S., 175, 187, 196
Goldiamond, I., 177, 187
Gollin, E. S., 177, 187
Goodenough, D. R., 169, 198
Goodnow, J. J., 37, 45, 179, 183, 187
Goodwin, W., 156, 192
Gordon, William J. J., viii, 7, 16, 103, 121–122, 124–125, 127, 132, 188
Gough, H. G., 167
Green, C. C., 70, 188
Gregg, L. W., 69, 188
Grove, R. B., 70, 193
Guilford, J. P., 135, 151, 166, 167, 168, 169, 170, 171, 172, 182, 183, 188
Guilford tests, 99, 166

Haag, R., 59, 193
habit, in problem solving, 10, 18–19, 24–25, 33, 34–36, 56, 161
habit hierarchies, 5, 6, 40, 47–53, 72
Hadamard, Jacques, 16, 20, 188
Harbridge House, Inc., 7
Harding, H. F., 25, 193
hatrack problem, 35, 47, 176
Hayes, J. R., 178, 188
Haygood, R. C., 45, 188
Henle, M., 179, 189
heredity, and problem solving, 18
Hess, J., 175, 184

heuristic computer programs, 61–62, 65, 67, 70, 162–163
Hoffman, L. R., 13, 189
Holder, J. R., 179, 187
Holland, J. L., 173, 189
Houtman, Susan E., 106, 108, 111, 112, 113, 114, 126, 135, 136, 138, 139, 140, 147, 148, 149, 173, 185, 189
Hovland, C. I., 70, 189
How to Make More Money (Small), 110, 111
How to "Think Up" (Osborn), 7
Huff, E. M., 179, 183
Hull, Clark L., 5, 48, 49–50, 56, 189
Hunt, E. B., vii, 45, 60, 62, 67, 70, 71, 73, 74, 189
Hyman, R., 171, 189

I Am the Mayor task, 83–84, 177
Idea Books, 150, 151–153
idea checklists, 8, 9, 76, 103, 107–108, 118–119, 136, 139, 140, 148, 163; Arnold, 109; Davis-Houtman, 112; Mason, 115; Osborn, 108–109; Polya, 117–119; seven-item, 113; Small, 110–111; U.S. Air Force, 116; U.S. Navy, 116
idea creation, vs. evaluation, 17
idea-squelchers, 19
illumination, in problem solving, 24, 27
in-basket problems, 81–83, 83–84, 176, 177, 179
incubation, in problem solving, 15, 16, 17
inquiry activity, 8, 75, 80, 83–84, 85–86, 88–89, 163; episode analysis, 86–88; teacher's in-basket task, 81–83
insight, 27, 28, 30, 31, 32, 39, 65, 73, 162, 163
inspiration, in problem solving, 15, 16, 24
intuition, vs. learning, 17–18

intuitive leap, 9
Invention Design Group, 122

Jaasted, K., 101, 186
Jackson, P. W., 168, 169, 172, 187
Johnson, D. M., vii, 170, 189
Johnson, E. S., 70, 190
Johnson, W., 175, 178, 181
Jones, H. L., 84, 183
Judson, A., 176, 190

Kaplan, I. T., 177, 190
Karp, S. A., 169, 198
Katona, G., vii, 33, 170, 175, 177, 190
Kendler, H. H., and T. S., 46, 56, 57, 177, 190
Kessler, D. K., 175, 187
Kilpatrick, J., 177, 190
Kingsley, H. L., 16, 190
Klausmeier, H. J., 156, 193
Kogan, N., viii, 134, 165, 169, 171, 197
Köhler, Wolfgang, vii, 4, 5, 6, 12, 18, 27, 29, 30–33, 39, 41, 43, 49, 56, 59, 162, 163, 176, 190
Koos, E. M., 80, 85–86, 181, 183, 190
Kotovsky, K., 70, 177, 195

letter-combination hierarchies, 51
Lewis, D. J., 188
Lewis, N., 192
Linker, E., 175, 190
Little, Arthur D., Inc., 122
Logic Theorist (LT), 64–65, 66, 67, 69, 71–72
Lorge, I., 177, 190
Loupe, M. J., 79, 80, 194
Luchins, A. S., vii, 27–28, 29, 33–34, 39, 47, 58, 162, 179, 190, 191
Luchins, E. H., vii, 179, 191
Lyda, L., 176, 191

MacKinnon, W. J., 175, 179, 198
Maier, N. R. F., vii, 28, 29, 33,

Index

34-35, 39, 47, 51, 162, 175, 176, 178, 179, 191
Maltzman, I., 49-50, 55, 59, 191
Manske, M. E., 51-52, 94, 102, 134, 165, 185, 191
Mari, S. K., 140, 149, 185
Marks, M. R., 178, 191
Mason, J. G., 16, 92, 95, 107, 109, 115, 116, 191
Mater program, 68
Mayzner, M. S., 51, 191
mechanistic, vs. mentalistic psychology, 27, 29
Mednick, S. A., 166
Mentality of Apes (Köhler), 30
Merrifield, P. R., 167, 168, 170, 171, 173, 188, 191
metaphorical thinking, 120, 121, 122, 128, 135, 136, 139, 163
Miller, G. A., 69, 191
Moore, O. J., 179, 191
Morgan, C. Lloyd, 41
Morphological Astronomy (Zwicky), 105
morphological synthesis, 76, 103, 105-107, 136, 139, 140, 163
Moss, G., 190
Myers, R. E., 150, 151, 153, 156, 157, 164, 192, 197

Needham, W. E., 170, 183
Newell, A., vii, 28, 63, 64, 65, 66, 67, 71, 73, 74, 177, 192, 195
Niemark, E. D., 192
novel sentences, 54-55

Odom, R. D., 46, 178, 192, 195
Olton, R. M., 156, 184, 192
operant conditioning, 6, 40, 53-56, 72
O'Reilly, J., 59, 193
Osborn, A. F., viii, 6-7, 16, 17, 92, 94, 95, 96, 101, 102, 103, 107, 108, 109, 113, 114, 126, 135, 193

Paige, J. M., 175, 193
Paivio, A., 74, 193
Papanek, V. J., 128, 130, 193

Parnes, S. J., viii, 25, 94, 107, 109, 193
part-changing method, 139, 148
Pavlovian conditioning, 57
pea-transfer problem, 47, 175, 177
pendulum problems, 20, 22, 35, 47, 49, 51, 177-178
Perceiver, 67-69
perceptual reorganization, 5, 27
personal analogy, in synectics, 122-125, 127
Pettigrew, T., 179, 187
Phillips 66 technique, 96
Picasso, Pablo, 17
Piper, R. M., 79, 80, 194
Plots, Puzzles and Ploys (Idea Book), 152
Polya, G., 16, 117-118, 177, 193
Poznyanskaya, E. D., 68, 197
predisposition, and problem solving, 18
preparation, and problem solving, 15, 24
Prince G., 122, 123, 124, 126, 127, 193
Principia Mathematica (Whiteside and Russell), 64, 71
probability-learning tasks, 45-47
problem, defined, 12-13, 21, 24, 48; solution, defined, 14-15, 24
problem solving: abilities, 157, 164; attitudes, 8-9, 102, 134-135, 136, 157; cognitive restructuring, 27, 30, 162; constructive discontent, 135; defined, 32-33, 161; dimensions of, 21-24; direction, 35, 73, 162, 163; divergent vs. convergent, 22, 25, 50, 53; fixation, 5, 18, 19, 24, 25, 35-36, 48, 57, 73; habit and conformity, 10, 18-19, 24-25, 33, 34-36, 56, 161; habit hierarchies, 5, 6, 40, 47-53, 72; heredity, 18; incubation, 15, 16, 17; illumination, 24, 27; insight, 27, 28, 30, 31, 32, 39, 65, 73, 162,

problem solving (*cont'd*)
163; inspiration, 15, 16, 24; intuition vs. learning, 17–18; intuitive leap, 9; metaphorical thinking, 120, 121, 122, 128, 135, 136, 139, 163; perceptual reorganization, 5, 27; predisposition, 18; preparation, 15, 24; recentering, 27, 28, 36, 162, 163; rigidity, 18, 25, 162; set, 5, 18, 25, 65; steps and stages in, 15–17, 24, 79–80, 126–128, 139–140, 161; task variables, 4–5; taxonomy of, 20–24, 25, 161; theories of, 5–6, 10, 58–59, 72–73; tradition-orientation, 18, 25; training in, 6–8, 24; transfer, 6, 57–58, 72, 73, 162; trial-and-error, 6, 21–22, 23, 24, 28, 40–43, 43–47, 59, 62, 65, 72, 162; as voluntary act, 133–134

product improvement problems, 22, 170

Productive Thinking Program, 150, 155–156

programming, in computer simulation, 62–64

Pryor, K. W., 59, 193

psychology: cognitive theories, 27–28, 29, 39, 72–73; Gestalt, 5, 6, 18–19, 27–28, 29, 39, 40, 73, 80, 88, 162; mechanistic vs. mentalistic, 27, 29; S-R theory, 27–28, 29, 39, 40, 53–55, 56–57, 58–59, 72, 73, 80, 162

Purdue Creativity Program, 150, 153–155, 164

puzzle box, 41, 42, 48, 49, 53–54, 59

Raaheim, K., 47, 177, 193, 194
Ray, W. S., 176, 179, 193
Raytheon, 130
recentering, 27, 28, 36, 162, 163
reference ego, 81
Reitman, W., 70, 193
response chaining, 6, 55

response hierarchies, 162
Restle, F., 45, 47, 49, 176, 183, 193
Richards, J. M., 170, 183
rigidity, 18, 25, 162
rings and peg problem, 36
Ripple, R. E., 156, 197
Robinson, J. A., 70, 194
Robinson, W. L. T., 155, 194
Rohwer, W. D., Jr., 74, 194
Ronda, T., 156, 193
Ronning, R. R., 5, 51, 194
Rookey, T. J., 173, 194
Ross, B. M., 194
roundabout-way tests, 30
Roweton, W. E., 102, 113, 114, 126, 140, 149, 173, 185
rule-out theory, 5
Russell, Bertrand, 64, 65, 71, 72, 197

Saint program, 70
Saturday Subway Ride (Wisconsin), 136, 141–147
Saugstad, P., 47, 194
Saunders, D. R., 89, 187
Schaefer, C. E., 167
Schlosberg, H., 41–42, 197
Schulz, R. W., 57–58, 194
Scott, J. A., 11, 132, 185
serendipity, 57
serial-position curve, in learning, 51–53
series problems, 20, 176, 177
set, 5, 18, 25, 65
Shaw, J. C., vii, 28, 63, 64, 65, 71, 74, 192
Shaw, M. E., 194
Sheerer, Martin, 28, 29, 33, 35–36, 39, 162, 163, 194
Shirk, E. J., 177, 187
Shoup, R. G., 70, 193
Shulman, L. S., 13, 79, 80–83, 84, 163, 194
Simberg, A. L., 107, 119, 195
Simon, H. A., vii, 28, 63, 64, 65, 66, 67, 68, 69, 70, 71, 73, 74, 175, 177, 182, 188, 192, 193, 195
Simon, P., 195

Skinner, B. F., 12–13, 28, 53–56, 195
Slagle, J. R., 70, 195
Small, Marvin, 110–111
Society of American Value Engineers, 7
Spiegel, J., 190
Staats, Arthur W., 28, 54–56, 195
Staats, C. K., 54, 195
Stein, L. S., 195
steps, in problem solving, 15–17, 24, 79–80, 126, 139–140, 161
Stevenson, H. W., 45–46, 195
stick and banana problem, 31, 32, 39, 43, 178
stimulus-response (S-R) psychology, 27–28, 29, 39, 40, 53–55, 58–59, 72, 73, 80, 162; vertical vs. horizontal processes, 56–57
stop-and-go brainstorming, 96
Stothart, J. R., 85, 183
Study of Thinking (Bruner) 37, 38
Suchman, J. R., 80, 84, 86, 87–88, 163, 196
Suedfeld, P., 175, 196
Sultan, 12, 18, 31–32, 43
switch-light problems, 20, 22, 23, 24, 43–44, 46
Sylvania Electric Co., 77
synectics, viii, 8, 9, 76, 135, 136, 139, 163; analogy strategies, 122–125, 127; force fit, 127; Gordon's contributions, 121–122; leadership function, 125–126; problem-solving stages, 126–128
Synectics (Gordon), 103
Synectics, Inc., 7, 121, 122

Tab Inventory of Science Progress, 84–85
Taft, R., 167, 173, 193
task variables, 4–5
Taylor, C. W., 25, 171, 196
Taylor, D. W., 101, 179, 196
taxonomy, of problem solving, 20–24, 25, 161

teacher's in-basket task, 81–83, 83–84, 176, 177, 179
Techniques of Creative Thinking (Crawford), 103
Thinking Creatively (Davis-Houtman), 106, 107, 111–113, 136–140, 141, 147
Thorndike, E. L., vii, 4, 6, 28, 40–41, 42, 48, 49, 53, 59, 178, 196
Thurstone, L. L., 167, 169, 196
Tichomirov, O. K., 68, 197
Tolman, E. C., 44
Torrance, E. P., viii, 9, 99, 150, 151, 153, 155, 156, 157, 164, 168, 170, 172, 173, 192, 197
Torrance Tests of Creative Thinking, 99, 155, 166
Trabasso, T., 45, 183
tradition-orientation, 18, 25
Train, A. J., 113, 114, 185, 197
transfer, 6, 57–58, 72, 73, 162
Treffinger, D. J., 150, 153, 154, 155, 156, 157, 186, 197
Tresselt, M. E., 51, 191
trial-and-error, 6, 32–33, 34, 35, 38, 40–43, 59, 62, 65, 72, 162; in computer simulation, 62, 65; implicit vs. overt, 43–47
Tuckman, J., 190
two-string problem, 47, 51, 56, 179

unusual-uses problems, 22, 79, 172

value engineering, 7
Value Engineering Department, Lockheed-Georgia, 9, 11, 197
verification, 15, 24
Vernon, J., 175, 196

Wake Up Your Mind (Osborn), 7
Wallach, M. A., viii, 134, 165, 169, 171, 197
Wallas, G., 15, 127, 197
Wand, B., 89, 187
Wardrop, J., 156, 192

Warren, T. F., 100, 102, 113, 114, 119, 126, 140, 149, 173, 185, 197
Wason, P. C., 197
water-jar problems, 19, 20, 22, 33–35, 47, 58, 162, 179
WBAA School of the Air, 150, 197
Weir, M. W., 46, 195
Weisskopf-Joelson, E., 101, 197
Welch, L., 172, 198
Wertheimer, M., vii, 33, 198
Whitehead, A. N., 64, 65, 71, 72, 197
Wiener, M., 175, 184
Willis, J. E., 175, 179, 198
Wilson, R. C., 188
Wisconsin Project, 77, 133–149
Wisconsin Research and Development Center for Cognitive Learning, ix
Witkin, H. A., 169, 176, 198
Wittrock, M. C., 176, 198
Woodworth, R. S., 41–42, 198
Write? Right! (Houtman), 147–148

X-ray problems, 20, 22, 179

Your Creative Power (Osborn), 7

Zagona, S. V., 175, 179, 198
Zand, D. E., 179, 198
Zerbolio, D. J., 170, 189
Zimmerman-Guilford Interest Inventory, 167
Zwicky, F., 105, 119, 197